# The Bitcoin Shortcut

## Quickly Learn How Bitcoin Works, How to Invest, and Why It's Changing the World

*Craig Anthony, CFA*

Bitcoin Strategic Solutions LLC
www.btcstrategic.com

The Bitcoin Shortcut: Quickly Learn How Bitcoin Works, How to Invest, and Why It's Changing the World

ISBN: 978-0-9903202-8-9 (hardcover)
ISBN: 978-0-9903202-5-8 (paperback)
ISBN: 978-0-9903202-3-4 (eBook)

First Edition
v1.9

Published by Bitcoin Strategic Solutions LLC.

Printed in the United States of America.

*This book is dedicated to my amazing wife*
*Emiko and our beautiful daughter Aiko.*

Bitcoin will change your life.

# CONTENTS

## Part 4
## *Investing in Bitcoin*

# ABOUT THE AUTHOR

**Craig Anthony, CFA** is a Bitcoin consultant, analyst, and software developer. He writes about Bitcoin at www.craiganthony.io and is currently focused on building his company Bitcoin Strategic Solutions LLC. His background includes tech startups and professional investment management.

## Is it Bitcoin or bitcoin?

The word Bitcoin refers to two things. When capitalized, Bitcoin refers to the software, the network, and the concept of Bitcoin. Such as, "The world is better because of Bitcoin."

When a lowercase "b" is used, bitcoin refers to the monetary units within the Bitcoin network. For example, "She sent him 1 bitcoin." These monetary units are also commonly referred to as BTC on cryptocurrency exchanges.

Throughout the book, we'll use Bitcoin, bitcoin, and BTC.

# What is The Bitcoin Shortcut?

*The Bitcoin Shortcut* is the most thorough, beginner-friendly resource on Bitcoin available on the market today. It teaches you what Bitcoin is, exactly how it works, and how you can start investing in it today.

It's written for everyday people yet doesn't avoid the nitty-gritty details. Bitcoin is a complicated and fascinating topic, so we do go down the rabbit hole. But I keep things clear and simple for you. I try to leave no question unanswered that a beginner might have.

Why is it a shortcut?

I put in roughly 1,000 hours studying Bitcoin and sifting through an overwhelming amount of material, so you don't have to. During my research, I read dozens of books, watched hours upon hours of video, read countless online articles, and tinkered with actual lines of code. Then, I synthesized all this information with my decades of professional experience (software engineering and investments) and my academic finance training.

The result is *The Bitcoin Shortcut.* It's the book I wish I had when I first started learning about Bitcoin. **Once you read it, you'll know Bitcoin better than 99.9% of people and be able to confidently start your Bitcoin investing journey.**

In summary, *The Bitcoin Shortcut* is the quickest path to deeply understanding Bitcoin.

# INTRODUCTION

Bitcoin is changing the world. There are about 160 million people across the globe that currently own it, with 2 million more being added to that each week. In only 12 years, Bitcoin's total market value surpassed the $1 trillion mark. That's about four times faster than Apple and twice as fast as Amazon or Google. The momentum behind Bitcoin is impossible to deny.

But what *is* Bitcoin?

Bitcoin is three things. It's a revolutionary new form of money, a digital monetary network, and a digital asset. It brings together ideas from many fields, including economics, finance, law, politics, computer science, and philosophy. There's never been anything like it in history.

Without question, Bitcoin is the most important financial technology ever created. It has the potential to objectively improve the lives of billions of people around the world and is already well on its way to doing so.

Why?

Bitcoin enables a complete shift in the way that we transact and store value. Anyone on the network can securely save money, make instant payments, and transfer value anywhere in the world at the speed of light. Incredibly, Bitcoin does this without the need of any trusted third-party intermediaries. Though long thought technically impossible, Bitcoin's peer-to-peer network eliminates the need for banks, payment processors, money agents, and similar financial middlemen.

But there's much more to Bitcoin.

Unlike traditional financial institutions, Bitcoin has a high level of privacy and users have *total control* over their funds. Payments can't be censored and tracing them to a specific individual can be complex or even impossible. Money on the network can't be seized, frozen, or stolen. Bitcoin is beyond the reach of companies, banks, governments, dictators, and thieves.

Bitcoin has no central authority. It's decentralized, spread across tens of thousands of computers all over the globe. It's not a company or a platform—it's a protocol. No single person, entity, or country owns the network. And no single person, entity, or country can control, manipulate, or influence the network's operations.

Bitcoin was designed and engineered with safety, durability, and security in mind. Monetary value is stored on the network and seamlessly moved between people every day. Ownership and transaction records are kept with flawless integrity.

Yet, the need for any two Bitcoin users to trust each other at all during a transaction is eliminated—this seems counterintuitive and unworkable, but Bitcoin does it and does it brilliantly. At the network level there has never been a hack and not a single bitcoin has ever been stolen since Bitcoin launched in 2009.

Bitcoin is unstoppable. It's changing the world and will change *your* life. Now it's your turn to understand, embrace, and invest in Bitcoin like millions of others. There's never been a better time. So, what are you waiting for? Let me show you the beauty, the power, and the potential of Bitcoin. It's a once in a lifetime opportunity you shouldn't miss.

Craig Anthony, CFA
New York City, July 2021

# Part 1

## *Preliminaries*

*Preparing you to learn about Bitcoin. All the important things about money you probably never thought about. A brief description of digital money. Intermediaries and ledger madness. The first hints of why Bitcoin is such an incredible technology.*

CHAPTER 1

# MONEY

There's *much* more to money than meets the eye. Those bills in your wallet and coins in your pocket aren't just cotton, linen, and base metals. They're the physical form of some amazing concepts you've likely never even thought about.

Since Bitcoin is a new form of money and a giant leap in monetary technology, learning about money itself is the logical first step for anyone starting their Bitcoin journey. Only with this foundational knowledge of money will you be able to fully appreciate Bitcoin and its role in changing the future of finance and investments.

While it may seem mundane to read an entire chapter about money, I promise you that it's both interesting and important. The design and engineering of Bitcoin tackles everything discussed in this chapter, so keep that in mind as you read.

Let's get started.

## MONEY IS A TECHNOLOGY

It's easy to forget that money is a technology that has evolved and improved over time.

We know that around 11,000 years ago livestock and grains were used as money. Banking originated in Babylon approximately 4,000 years ago and around that same time standardized units of silver started being used as money.

The first known coins were used as money about 2,700 years ago. In Egypt, around 2,300 years ago, warehouse receipts were used for the first time to transfer value between accounts. Paper currency was first developed by the Chinese roughly 1,200 years ago.

The earliest known foreign exchange contract was in the year 1156, with payment due in Constantinople. In 1792, the Dollar was adopted in the United States because of the US Coinage Act.

The first universal credit card was introduced in 1950 by the Diners' Club Inc and 28 years later The First National Bank of Seattle issued the first debit card. PayPal, established in 1998 under the name Confinity, combined money and email making online payments as easy as clicking a few buttons.

But perhaps the most important dates in the history of money are October 31st, 2008, and January 3rd, 2009. It was on that October day that Bitcoin was introduced to the world by its creator, Satoshi Nakamoto, through a post on an online forum called The Cryptography Mailing List. He wrote: "I've been working on a new electronic cash system that's fully peer-to-peer, with no trusted third party." He included a link to his revolutionary whitepaper.

On the following January 3rd the Bitcoin network came into existence as Satoshi mined the first block of the Bitcoin blockchain, creating the very first 50 bitcoin. At that moment cryptocurrency was born and money took on a brand-new form. The future, a world of fully decentralized global finance, will be able to trace its roots back to this day.

How did we get here? To begin to understand this we must start by covering different forms of money and dig deeper into what problems money attempts to solve.

## DIRECT EXCHANGE

The *direct exchange* of valuable goods or services with other people is known as barter. This type of value exchange has always existed in human society and still does today all over the globe. When all the parties involved in a trade agree on its terms barter works great. You have five apples that I want, and I have three avocados that you want. Let's trade. Deal.

But there is a big problem that arises with barter that's known as the problem of *coincidence of wants*. It occurs when what you want is produced by someone who does not want what you have. For example, you want my avocados, but I don't want your apples. Maybe I'm only willing to part with my delicious avocados for beef jerky.

Now you must go find someone who produces beef jerky *and* is prepared to trade it to you for your apples. Then you can come back to me, and we'll do business trading beef jerky for avocados. Or I can go find that beef jerky producer myself and hope that he'd trade me some for my avocados. If he hates avocados, I'm out of luck.

Beyond this basic inefficiency of direct exchange, three specific aspects of the problem of coincidence make things even more inconvenient. They are coincidence of location, coincidence of scales, and coincidence of time frames.

First, a lack of coincidence of location exists when the objects meant for barter aren't in the same physical place. To make matters potentially worse, one or more of the objects involved in the barter might even be immovable (houses for example).

Second, a lack of coincidence of scales means that what you want to trade is not equally valuable as what you want to trade it for and dividing the items into smaller units may be impractical or impossible.

Try using apples to trade for a new car. The first problem is that even if you had enough apples, the owner of the car probably doesn't want that many apples at once. Second, if you tried to buy a car one apple at a time over many decades, the car would have to be broken down into such small pieces that it likely couldn't ever be assembled again while maintaining its functionality.

Third, a lack of coincidence of time frames is a problem when one side of the barter wants to trade perishable goods for non-perishable goods. Trying to amass enough apples to buy your new car in time before they all rot to nothingness would be difficult. Nobody wants shriveled, moldy apples.

In small groups of people with a very limited number of goods produced barter can work. But as an economy grows in population size and more people specialize in producing different goods, barter becomes totally impractical. People would need to perform multiple trades, perhaps even hundreds, just to obtain a single item they want or need.

The way this avoid this trade-o-rama is to use *indirect exchange.*

## INDIRECT EXCHANGE

*Indirect exchange* beats barter hands down. It's a two-step exchange process. Step one is to exchange your goods or services for an intermediary good. Step two is to exchange the intermediary good for the good or service that you initially wanted. The intermediary good is known as a *medium of exchange.* Just about anything can be used as a medium of exchange if enough people agree on it.

But as an economy grows, division of labor increases, and more goods and services become available for trade, a new problem emerges. There's a possibility that different groups within the economy use different mediums of exchange. While the hideous trade-o-rama of barter would be avoided, an impractical number of mediums of exchange among counterparties could still exist. The need for a single agreed upon intermediary good arises.

When an intermediary good becomes a medium of exchange that's widely accepted it's called *money*. The good does not have to be universally accepted, just widely accepted. Not even the most powerful fiat currency in the world, the U.S. dollar, is universally accepted.

Lots of different items have been and are used by humans as money. Cattle, seashells, stones, glass beads, animal skins, silver, gold, salt, tobacco, and government paper are somewhat normal examples. But some items used as money can be quite strange. The regional bank Credito Emiliano in Italy accepts Parmigiano Reggiano cheese wheels for collateral on certain loans. Laundry detergent is used as money by some drug dealers. Even urine is sometimes used as money in prisons.

But how is money chosen? Who decides what the intermediary good should be? There are many factors that determine if an intermediary good should be used as money. Let's have a look at some.

## SALABILITY

The term *salability* (sometimes spelled saleability) refers to the degree of ease with which a good can be sold for a reasonable economic price on the market whenever the owner desires. Any good can be used as a medium of exchange but different goods have different levels of salability.

A good's relative level of salability can be analyzed by revisiting the problem of the lack of coincidence of wants. How salable is the good across all three dimensions of the problem: location, scales, and time?

- *Salability across location* refers to the ease in which the good can be transported
- *Salability across scales* means that the good is easily divisible into smaller units or grouped into larger units
- *Salability across time* indicates that the good will maintain its value in the future

Salability across time is the most important of the three because it enables people to build wealth. When wealth can be stored in a particular good that good is known as a *store of value.*

For a good to be salable across time it can't be susceptible to deterioration, decay, or rot. Milk or apples are not salable across time because they turn nasty quickly. You couldn't build wealth using jugs of milk or baskets of apples. Gold on the other hand is highly salable across time. You could build wealth in gold over time without worrying about it turning to junk.

## HARD MONEY, EASY MONEY, AND STOCK-TO-FLOW

If a particular form of money has a supply of units that's difficult to increase it's called *hard money*. On the other hand, if a particular money's supply of units can easily and effortlessly be increased it's called *easy money*. To put it another way, hard money can't be created easily because it requires a large amount of effort to be produced. Easy money can be created easily, sometimes even effortlessly.

Gold is considered hard money for example, because the process of finding it, extracting it from the Earth, and refining it is difficult, expensive, and time consuming. Government currency, on the other hand, is easy money. Politicians could quickly whip up new legislation that results in trillions of currency units being printed in no time at all.

The hardness of money is important. If the supply of money increases dramatically over the period of time you own it, it's going to lose value and purchasing power. Hard money is a great store of value. Easy money is not.

How can we tell if a certain type of money is hard or easy? And how can we determine its relative hardness? There's a metric called the *stock-to-flow* ratio (sometimes called SF or S2F) that can help. The stock of a good is the current existing supply of the good. This is total that has ever been produced, less any portion that was consumed, destroyed, or lost. The flow of the good is the new supply produced in the next given time period, usually measured annually. The stock-to-flow is the ratio between these two numbers.

The *higher* the stock-to-flow ratio, the *less* new supply of the good enters the market each year relative to the existing supply. A higher stock-to-flow indicates more hardness and the likelihood that the good will retain value over time. In contrast, goods with low stock-to-flow are not likely to retain value over time since new supply is generally used only to replenish consumed inventory—production keeps up with demand.

For a good to be an excellent store of value and more salable across time it must be hard. Otherwise, if people started using it as money, producers would focus much more of their resources on it and the flow would become much larger relative to supply. This would lower prices, devalue the good and cause wealth stored in the good to evaporate.

## ACCEPTABILITY

If the intermediary good in question is highly salable and has a high stock-to-flow ratio it's more likely to be accepted by a large number of people as a money. With wide enough acceptance, the good can then be used as a *unit of account*, enabling a meaningful interpretation and understanding of the price of that good relative to a variety of assets.

If you live in the United States just about everything is accounted for in U.S. dollars. The only reason you don't say something like "I have 75 baseballs worth of food in the fridge" is because baseballs are not a widely accepted unit of account. People would think you're crazy.

## FUNCTIONS AND CHARACTERISTICS OF MONEY

Let's sum up what you've learned so far. For something to be categorized as money the consensus is that it must serve the following three major functions:

- Be a **medium of exchange.** It must be a widely accepted payment mechanism used for buying goods and services and paying off financial obligations.
- Be a **store of value.** It must be worth relatively the same today as it is tomorrow, next week, next year, and so on (and purchase roughly the same amount of goods and services over time).
- Be a **unit of account**, meaning you can use it to compare the value of two or more items, confidently

calculate the value of assets, and do accurate and meaningful accounting with it.

Useful money must also be highly salable and have the following five characteristics:

- *Durability* - have long-lasting physical integrity
- *Divisibility* - ability to be divided into smaller units
- *Portability* - easily transportable from one to location to another
- *Uniformity* - units of the same denomination must have the same purchasing power
- *Fungibility* - one unit can be interchanged for another of like kind

And the harder the money, the better.

## CATEGORIES OF MONEY

Now that you know enough about the concept of money it'll be helpful to the learn about the three major categories of money. They are:

- *Commodity money* - has value because the substance it's made of has value
- *Representative money* - a certificate or claim on an underlying commodity
- *Fiat money* - gets its value from government authoritative order and is not backed by any asset

What about Bitcoin and cryptocurrency? We'll get there soon enough. For now, let's focus on the three commonly accepted kinds of money.

## Commodity Money

Commodity money has *intrinsic value*, meaning that the stuff the money is made of is valuable and useable outside of its function as money. This intrinsic value is directly identifiable by its users—the utility or beauty of the object is obvious. Salt, for example, can be used to season or preserve food. Gold and silver can both be used to make jewelry or in industry. No matter what the object is that's being used as commodity money, its value outside of being money is clear.

Besides salt, gold, and silver, some other examples of goods used as commodity money include cattle, copper, tea, silk, cigarettes, cocoa beans, and grains. Commodity money has been used for thousands of years.

## Representative Money

Representative money can be certificates, receipts, or tokens that are claims on an underlying commodity or asset. Representative money can be exchanged for the underlying item either on demand or sometimes on a specific date. This type of money must have something of value "backing" its face value. While the money represents value, it has no intrinsic value of its own.

Here's an example. Rather than carrying around a chest of gold all over town to buy something you could store that gold in a bank vault. The bank would then give you some paper or certificate representing the gold bars and your ownership of them. Carrying around this document from the bank is much easier and safer than transporting the actual gold.

But representative money introduces *counterparty risk*. That's the risk that a counterparty involved in a transaction might default on their obligation. In the example above, the bank issuing the representative money is the counterparty.

While it's a low probability of happening, there's always a chance that the counterparty can't or won't honor the document's claim on demand.

Over time as people began to trust representative money just as much as the commodities themselves, it replaced commodity money. Today, in most cases around the world, representative money has been replaced with fiat money.

## Fiat Money

Fiat money does not have any intrinsic value, nor does it represent a claim on any asset. It's completely de-linked from anything valuable and is only considered money because a country's lawmakers say it is. Legislation declares fiat money *legal tender*, meaning that it must be accepted as payment in the jurisdictions the law covers.

So how does fiat money get its value? It is a combination of supply and demand, the stability of the issuing government, the condition of the economy in which its used, and the faith (in it) of the people who use it. Fiat currency gives central banks a large degree of control over the economy since they can easily manipulate the supply of money, expanding or contracting it at will.

One advantage of fiat money is its relative stability. Governments have more flexibility in responding to economic variables and events compared to commodity money, however when taken too far, this can cause troublesome inflation. Fiat money also has excellent seigniorage, which is difference between the face value of the money and the cost of producing it. This difference accrues to the government as economic profit. Seigniorage isn't always positive though. For example, the U.S. penny and nickel both currently cost more to produce than their face value.

Almost every major sovereign currency in the world today is a fiat currency.

## A Comparison of Money

Let's look at how these three forms of money compare to each other. Some of these metrics we have not discussed yet but will in subsequent chapters. In this comparison we'll use gold since it's the world's number one commodity money.

In addition to the five characteristics we already discussed, six more metrics are added: verifiability, scarcity, censorship resistance, decentralization, unforgeable costliness, and programmability.

For all metrics on the table, the higher the score the better the money.

- *Verifiability* - The ability of a unit of the money to be authenticated and how trustworthy that demonstrated authentication is
- *Scarcity* - The rarity of the money
- *Censorship Resistance* - The degree to which money is tamper-proof; the degree to which transactions can't be prevented, stopped, reversed, or modified
- *Decentralization* - The degree to which planning, decision making, and control is spread across location and group, and not concentrated
- *Unforgeable Costliness* - The degree to which the cost and effort of producing the money can't be faked or spoofed
- *Programmability* - The ability to accept change, to accept new sets of instructions to alter behavior

In Chapter 4 we'll add Bitcoin to the following table, and you'll see how it dominates.

## Money Metrics Table

| Metric | Gold | Rep. | Fiat |
|---|---|---|---|
| Durability | H | L | L |
| Divisibility | L | M | M |
| Portability | L | H | H |
| Uniformity | H | H | H |
| Fungibility | H | H | H |
| Verifiability | M | M | M |
| Scarcity | M | M | L |
| Censorship Resistance | M | L | L |
| Decentralization | M | L | L |
| Unforgeable Costliness | H | L | L |
| Programmability | n/a | n/a | n/a |

Key: H = high, M = moderate, L = low

Note: Rep stands for Representative

## SOUND MONEY AND UNSOUND MONEY

*Sound money* is money that can hold value over time. It's not prone to any sudden appreciation or depreciation in purchasing power over the long term. Sound money also has the characteristic that it remains under control of its owner and protected from (perhaps government) coercion and meddling of the currency's system. The free market chooses what it considers to be sound money—no one person or entity gets to decide.

*Unsound money* is the opposite. It *is* prone to sudden appreciation or depreciation. Unsound money is not chosen by the free market to be money. Its supply is planned and controlled by central banks.

## TIME PREFERENCE

*Time preference* is the relative value you place on receiving something in the present versus receiving it in the future. Another way to look at it is the ratio of how you value the present compared to the future.

Not everybody has the same time preference. Say you lend me $100 today. Would you rather that I give you back $100 tomorrow, or $120 in thirty days? Your time preference changes depending on a variety of factors, some of which may be the level of trust between us, your financial condition, my financial condition, will either of us die before you get paid back, will the U.S. dollar collapse in thirty days, and so on. There are too many factors to list.

But the key idea is this: a lower time preference means that you prioritize future outcomes, while a higher time preference means that you prioritize present outcomes.

Time preference changes given the situation but knowing your time preference is a key aspect in making good decisions. Creating a better life for yourself requires making trade-offs and compromises with your future self. Placing more importance on your future well-being requires that you lower your time preference.

*Sound money enables all of us to lower our time preferences, save wealth (store value), and plan for the future more rationally without worry that our money will lose value over time.*

If you really want to dig into quantifying the concept of time preference, look up 'discount function' online. It's an economic model that will keep you busy.

* * *

We just unpacked a lot of important ideas about money. This will serve as the foundation upon which your Bitcoin knowledge can be built. Keep the ideas we just covered in mind as you progress through the book.

We'll discuss details soon, but for now just know that based on what this chapter covers Bitcoin blows fiat, representative, and commodity money out of the water. It was designed that way. Bitcoin is the latest iteration of money and it's the best.

You may not agree with those statements just yet but there's a high probability you will by the end of this book. There's so much more to uncover, learn, and digest. You'll be amazed at what Bitcoin is and what it's capable of. But we must take logical steps to get to the point where it all clicks and makes sense.

So, let's continue right away to the next (and last) chapter of Part 1. There are some ultra-important things about digital money that you need to understand before you can start learning all about Bitcoin.

CHAPTER 2

# DIGITAL MONEY

It's estimated that only about 8% of the world's money is physical money. The remaining 92% is digital money that's managed, stored, and exchanged on computer servers and systems.

Digital money is any form of payment or money that only exists electronically. If the payment or money is not a bank note, coin, or some other physical object accepted as money, it's digital money.

Credit cards, debit cards, bank wire transfers, ACH payments are widely used examples of digital money. The cash value transacted on Cash App, PayPal, Venmo, and similar payment apps falls into the category of digital money also.

Digital money makes transactions more convenient by not requiring the physical transfer of any bills, bars, or coins. This is especially useful over long distances. Plus, digital money transactions are usually completed quickly, sometimes about as close to instantaneously as you can get.

Bitcoin is a very special sub-type of digital money which should *not* be grouped in with the examples mentioned above. The technology behind Bitcoin is genius and is in a class of its own.

**Unless specifically noted, all mentions of digital money in this chapter refer to the regular, dull, electronic money and payment systems you're used to—not Bitcoin.**

## INTERMEDIARIES

While digital money helps solve some issues associated with physical cash and makes life more convenient, it has a major problem at its core: intermediaries. All forms of digital money require third-party intermediaries to keep accurate records of transactions and account balances.

If you want to use digital money, you're required to set up an account with the third-party intermediary that provides the digital money service. From that point on you have no choice but to trust that they'll keep accurate records of your money and perform the transactions you want (not censor you) in a timely manner.

Independent third parties in the financial world are typically regulated so you can rest assured that the account and transaction records they keep are accurate. As for actual transactions, there's not much you can do to guarantee that they'll be executed except keep your fingers crossed. Most transactions will go through no problem. But it's possible that you'll want to buy something from somewhere that the third party doesn't like or approve of. Your transaction could be blocked at their discretion.

This third-party censorship is not something to take lightly. Though it may be rare, it certainly happens. If you don't like the idea of someone telling you what you can and can't do with your money, you're not alone.

As for digital money transactions executing and clearing in a timely manner, most will. But that's not always the case either. Third-party payment processors must deal with fraud, theft, technical failures, hacks, and other issues that can delay execution and settlement. These issues also increase transaction costs and other fees.

Going through third parties always adds security weakness to your transaction and puts the possibility of surveillance on the table by third-party employees, political authorities, law enforcement, enemies, or just about anyone with the right computer skills. Third-party intermediaries always come with some level of counterparty risk, too.

Even given the awful problems and risks that exist, it's unfortunate to say that trusted third parties are absolutely necessary for traditional digital money payments. They prevent the *double-spending problem* from occurring.

## THE DOUBLE-SPENDING PROBLEM

Digital money is just a bunch of entries in an electronic database. Balances can easily be duplicated or falsified. The double-spending problem is the risk that a single unit of digital currency is spent more than once.

Imagine that you and I are using electronic money with each other, but we aren't using a bank as a middleman. We just keep our own records of how much money we have and what transactions we've done. Pretend your spreadsheet says you own 100 units of some digital money. Let's pretend it's dollars.

Before you make any payments to me, I want to see your spreadsheet so I can confirm you have the money. It's the prudent thing to do.

Say you're going to buy something from me for $500. But you only have $100. What's preventing you simply editing your spreadsheet to say you have $500? Nothing is preventing you. It's easy for you to simply change the entry from $100 to $500. Or, you could add four more $100 line items, bringing your total to $500. Either way, it's easy and no one could stop you from doing it. It's money from thin air.

I'd see your doctored spreadsheet and say "Oh, OK, you do have the money. Let's proceed with our transaction. Let's make entries on both of our spreadsheets saying that you paid me $500." Transaction done.

Imagine a different scenario. You have your actual $100 and decide you want to buy something from me for that $100. We do the transaction and add the entries to our spreadsheets. I confirm that you added a line that says, "Paid Craig $100 for gizmo. New Balance $0."

But after we part ways, you see that your friend Mike is also selling something that you want for $100. Since I'm gone, and no one else is around to see what you're doing, you delete the line item that says "Paid Craig $100 for gizmo. New Balance $0." Now you are back to a balance of $100 and pay Mike for the item he's selling.

It's clear that in both scenarios, you're spending money that you don't have. Nothing prevents anyone from editing digital records with a few keystrokes. If this was how we dealt with electronic money, all hell would break loose. The integrity of the entire global finance and banking system would be crippled.

If users of a digital money have no way of verifying whether units they've received have already been spent elsewhere, or if they are falsified units, the entire digital currency and system it runs on is worthless. This is where trusted, regulated third-party intermediaries are needed to oversee accounts and ensure integrity of the systems.

These third parties are often licensed by authorities and follow strict rules, regulations, and standards. And they themselves are watched closely by government regulators. You can have a high degree of confidence in them and the digital money they oversee for you. But this comes with the problems, risks, and fees we discussed earlier.

But guess what? Bitcoin solves the double-spending problem *without needing any third-party intermediaries.* That's right. Bitcoin removes trusted third parties from the process and is the first digital payment technology to do so. That's one of the reasons it's so revolutionary.

## LEDGERS

The way that third-party intermediaries keep track of account balances and payments is through *ledgers.* A ledger is a summary of transactions recorded by amount, account, and date. These days ledgers are almost always electronic entries in a database but there's nothing stopping a person or small business from using written pages in journals.

Every third-party intermediary maintains at least one ledger account for each user account of its service. To show you a simple example of how ledgers work let's pretend that you have an account at Luddite Bank, a fictitious bank in your town.

Let's say that you have $100 deposited with them. For you that $100 is an asset but for Luddite Bank it's a liability. It's a liability because they should give that $100 to you on demand if you walk in and ask for a withdrawal or should pay it to someone else if you instruct them to (for example paying a utility bill or sending money to friend after losing a bet).

Let's also assume you have a friend named Mike that also has an account at Luddite Bank with $100 in it. Here's how things look so far (transaction dates omitted):

**You**

| *Assets* | *Liabilities* |
|---|---|
| **Bank Acct $100** | |

**Luddite Bank**

| *Assets* | *Liabilities* |
|---|---|
| | **You $100** |
| | **Mike $100** |

**Mike**

| *Assets* | *Liabilities* |
|---|---|
| **Bank Acct $100** | |

Now, let's say that you pay Mike $25. Luddite Bank would make the adjustments to their liabilities, your assets, and Mike's assets as follows:

**You**

| *Assets* | *Liabilities* |
|---|---|
| **Bank Acct $75** | |

**Luddite Bank**

| *Assets* | *Liabilities* |
|---|---|
| | **You $75** |
| | **Mike $125** |

**Mike**

| *Assets* | *Liabilities* |
|---|---|
| **Bank Acct $125** | |

This is easy for the bank to do since it knows you have the money available and you and Mike both have accounts at the bank. But things quickly get more complicated when this isn't the case. Mike might use a different bank across town or across the country. He also might be living in a different country, using a bank there. Depending on where Mike is located, he might need payment in the same currency you're using or perhaps a totally different currency that maybe you never even heard of.

Multiple financial institutions are required to make all these different scenarios happen in the real world. Correspondent bank accounts, central bank accounts, and foreign exchange services are just some of the things that may be required to make your transaction go through (if they allow it and don't censor you).

Correspondent bank accounts are bank accounts that banks have with other banks. Central bank accounts are accounts that banks have with national banks that provide banking and financial services to a country's government and commercial banking system. And foreign exchange services specialize is trading one country's currency for another country's currency.

All these layers of institutions mean more transaction fees, more time to settle the transactions, more security risks, and of course, more ledgers to ensure everything is accurate and legit. The global banking system is a ledger bonanza.

But as you know, Bitcoin doesn't need third-party intermediaries. It avoids this layer cake of banking madness by utilizing a decentralized public ledger called *blockchain*. You'll be introduced to the basics of blockchain in Chapter 5 and you'll learn the technical details of it in Chapter 12.

* * *

In Part 1 of this book, you learned all the important things you need to know about money. The average person doesn't sit around thinking about salability, stock-to-flow, functions of money, time preference, financial intermediaries, double-spending, and so on. But now you can! And from now on, you'll be the life of every party you go to.

But seriously, all the stuff we covered in Part 1 is necessary for you to have a deep understanding of Bitcoin. Feel good that you learned it. It's foundational and it gives you an edge.

It's time to get into some interesting stuff. The next section of the book covers what a cryptocurrency is and gets your basic understanding of Bitcoin in motion.

Let's go to Part 2!

# Part 2

## *Bitcoin Basics*

*A high-level description of cryptocurrency. Two of my favorite implications of the technology. A few things on Bitcoin before we dig deep. Satoshi, supply, and a comparison. Components of Bitcoin. A brief description of how a transaction works.*

## CHAPTER 3

# CRYPTOCURRENCY EXPLAINED

Bitcoin is the world's first cryptocurrency. Since its invention, there are now over 4,000 cryptocurrencies in the world. Most of them have no following, no utility, no trading volume, and little to no market value. Most (if not all) are garbage in digital form, created by someone trying to get rich quick.

Bitcoin has enormous value and solves a very specific problem. For what it does, I have my doubts that any other cryptocurrency will ever replace it. Of course, other cryptocurrencies that solve *other* problems *might* become very popular and valuable. It's unclear but doubtful.

A simple yet obvious thing to always remember: Bitcoin is a cryptocurrency, but no other cryptocurrency is Bitcoin.

Take a second to just look at the word cryptocurrency. It's a big word that looks complicated. As John Oliver once put it: "Cryptocurrency—everything you don't understand about money combined with everything you don't understand about computers."

At this point you know enough about money that John's quote doesn't apply to you. As for the technology, the truth is that it *is* complicated. That's part of the reason I wrote this book. I wanted to simplify the main technological concepts behind Bitcoin so anyone could learn them.

For now, take solace in knowing that you don't need to understand the technical details to have a general grasp of what a cryptocurrency is, what it's doing, and what it enables. Technical details of how it works can come later.

If you zoom out and look at things from a high level, a cryptocurrency is:

> Digital Medium of Exchange + Public Ledger - Trust + Cryptography = Cryptocurrency

Start with a digital medium of exchange, add a public ledger, remove the need for trust between parties during transactions, use a few elements of cryptography to make things work, and you've got yourself a cryptocurrency. Here's a little bit about each component in that formula.

*Digital Medium of Exchange* - An electronic, widely accepted payment mechanism.

*Public Ledger* - A giant publicly available record book containing the details of all the transactions ever made with the cryptocurrency. New transactions are added all the time. This is the *blockchain* for that cryptocurrency.

*Remove Trust* - Digital transactions can now happen directly peer-to-peer without the need for trusted third parties or the need for trust between peers. The network verifies and guarantees the correctness of the data.

*Cryptography* - The practice and study of techniques to secure information within a communication so only the sender and intended recipient of a message can view its contents. Cryptography is also used for data integrity, entity authentication, data origin authentication, and other use cases. Different cryptocurrencies utilize different techniques from cryptography.

That's the general formula, though it's possible that different cryptocurrencies will add or remove variables.

## CHARACTERISTICS OF CRYPTOCURRENCY

Cryptocurrencies have their own set of characteristics that further set them apart from other forms of money as well as traditional banking and financial systems. Here are a handful of characteristics that any cryptocurrency worth its salt has.

### Decentralized

With traditional fiat currencies national central banks and authorities run the show. This is not the case with cryptocurrencies. There is no central authority controlling them. Transactions are processed and validated by a decentralized, distributed, open network that no single entity controls or influences.

Beyond being immune to political influence, the networks powering cryptocurrencies are also architecturally decentralized. This means that there's no single point of failure or vulnerability in the infrastructure.

Decentralized systems have high fault tolerance (less likely to accidentally fail), higher attack resistance (no central points, and a high level of security), and high collusion resistance (due to a required high level of consensus).

## Immutable

With your bank account, credit card, or PayPal if there's a fraudulent transaction, you trust that they'll fix it and give you your money back. If you accidentally send money to the wrong person there's a good chance the transaction can be reversed. If you see a surprise fee in your account, contacting customer support can likely make it disappear.

With cryptocurrency this can't happen. Since there's no trusted third parties involved to watch over and manipulate transactions, if necessary, all transaction records are made public and immutable (unchangeable).

Transactions are effectively written in stone on the cryptocurrency's blockchain. This immutability means that you need to take cryptocurrency seriously. You must pay careful attention. If you make a mistake sending someone some money and it goes to the wrong person, you won't get your money back. Luckily, there are tools and processes in Bitcoin that prevents this. You'll learn about these in Part 3.

## Anonymous (sort of)

It's a common misconception that all cryptocurrencies are totally anonymous. Most are pseudonymous. *With Bitcoin in particular, bitcoin addresses, private and public keys, and transactions are not directly tied to your real-life personal identity*. Nowhere is a transaction recorded as "Joe Smith from Williamsburg, Virginia sent 1 bitcoin to Sally Baker from Rachel, Nevada."

With that said, full privacy is possible but not guaranteed. It is possible to establish links between someone's identity and bitcoin addresses. Once an accurate association is made, tracking and analysis of transactions can be done. If you're careful and conscious of the need to be as private as possible with your cryptocurrency, you won't have any problems though.

Listen to the direct words of the creator of Bitcoin, Satoshi Nakamoto (more on him next chapter):

> "The possibility to be anonymous or pseudonymous relies on you not revealing any identifying information about yourself in connection with the bitcoin address you use. If you post your bitcoin address on the web, then you're associating that address and any transactions with it with the name you posted under. If you posted under a handle that you haven't associated with your real identity, then you're still pseudonymous."

Nodes in the Bitcoin network cannot tell which bitcoin addresses belong to which IP addresses. The Bitcoin blockchain does not store IP addresses. But since there are so many other cryptocurrencies out there you should do your research before transacting in them if privacy is something that concerns you.

### Algorithmic Supply and Potential Scarcity

Cryptocurrencies usually have a limited supply of monetary units. For instance, Bitcoin's money supply is hard-capped. Regardless if a particular cryptocurrency's money supply is capped or not, it's typically determined by an algorithm. Typically, units of the currency are released over time following a predetermined schedule. This is usually the case, but all cryptocurrencies are different. If you are going to invest in a particular cryptocurrency, make sure you understand how and when new units of the currency are created.

Programmed algorithmic money supply is an amazing

feature. No politician, government, bank, or influential person can affect the supply schedule of the cryptocurrency's units, ensuring no unknowns or surprises in total units available at any given time. No one can just "print" trillions of new units. The entire money supply and future growth of the money supply, at any given point in time, is publicly known and unchangeable. Depending on the cryptocurrency, this keeps the stock-to-flow under control and keeps the currency hard.

Limited supply makes cryptocurrencies deflationary in nature, meaning that their buying power increases over time. Plus, with limited supply any increase in demand or adoption naturally pushes prices higher, increasing the value of each individual unit. Cryptocurrency units can also be lost or destroyed, permanently reducing supply which has the effect of ever so slightly increasing the value of the remaining units.

On the other hand, fiat currencies theoretically have an unlimited supply. Central banks and governments can print as much fiat money as they want. Printing money out of thin air is simply not a good idea. Easy money is dangerous.

Fiat currencies are all fundamentally inflationary—their value decreases over time.

### Borderless

With no institutions in the way, cryptocurrencies are truly borderless. Think about how amazing this is. If you can connect to the cryptocurrency's network, you can move money to and from anywhere in the world and no one can stop you.

Traditional borderless payments involve layers of financial institutions that each must record and approve the transaction before moving it along. Each institution adds transaction fees and time delays to the movement of money.

Cryptocurrency payments are direct, person-to-person, with one usually low fee. And they get completed in seconds to minutes. (The archaic, clumsy, traditional financial system's days are numbered.)

A typical argument against this borderless feature of cryptocurrency is that it requires the internet and related infrastructure, so they can be "turned off" or "stopped" by grumpy politicians and governments. This argument fails. Bitcoin, for example, is a software protocol and not a dedicated hardware "network" in the traditional sense. It doesn't need massive infrastructure like the internet does and a high-speed connection is not required to use it. Any device that runs the Bitcoin software can communicate with others, and that communication itself is typically infrequent and small in size.

As Saifedean Ammous puts it in *The Bitcoin Standard*:

> "There are countless wired and wireless technologies for the transmission of data worldwide...in order to create a world in which no Bitcoin user is able to connect to other users, the kind of damage that would be needed to be done to the world's information, data, and connectivity infrastructure would be absolutely devastating."

In that situation we'd likely be dead already and not need to worry about using money.

## Antifragile

Not every cryptocurrency has this characteristic, but Bitcoin surely does. You can learn all about the idea of antifragility in Nassim Taleb's amazing book *Antifragile: Things that Gain from Disorder*.

Think of a fragile object, such as wine glass. It likes peace and order. A small shock can shatter it. A fragile object does not like stressors, volatility, or disorder. These things don't just weaken the fragile object, they destroy it. Fragile objects or ideas can't handle much before irreversibly breaking down.

Antifragile things *get stronger with stressors and volatility.* From the words of Nassim Taleb:

> "Some things benefit from shocks; they thrive and grow when exposed to volatility, randomness, disorder, and stressors and love adventure, risk, and uncertainty. Yet, in spite of the ubiquity of the phenomenon, there is no word for the exact opposite of fragile. Let us call it antifragile. Antifragility is beyond resilience or robustness. The resilient resists shocks and stays the same; the antifragile gets better."

Bitcoin gains strength from shocks. Its decentralized nature ensures that no one single person or entity can force changes on the network and yet it's an evolving system, improving through a global team of volunteer programmers. Any attempted attack on Bitcoin hardens the network as these programmers fix the software and propose the changes. If the majority of the network accept the changes, they take effect.

The increase in strength naturally leads to increased confidence of all users. Every thwarted attack leads to a stronger Bitcoin network than existed before the attack, and to a more confident userbase than before. Bitcoin has proven itself and hardened itself over more than a decade as evidenced by it not dying.

Bitcoin is also convex to errors, meaning that they get fixed rapidly. Any errors that occur off-chain, on cryptocurrency exchanges for example, provide quick teaching moments which in turn lead to security and user experience improvements that get implemented very fast.

As you continue reading this book, you'll discover even more reasons that make Bitcoin antifragile.

## IMPORTANT IMPLICATIONS OF BITCOIN AND CRYPTOCURRENCY

While there are many positive implications of cryptocurrency, there are two that I want to mention. If either of these hit home with you, you're going to love being a part of the Bitcoin community.

### Total Financial Control

One of Bitcoin's main value propositions is censorship-resistance. With Bitcoin, anyone can transact on the network if they follow rules of the network protocol. You're free to transact as you please and no one can stop you.

Traditional banking and payment solutions do not have this feature. When you swipe your debit card or tap the payment button on the screen of an app, you're hoping that the institution's algorithms and filters allow your transaction to execute.

You should be able use your money exactly as you want. But there's always a non-zero probability that a particular transaction won't go through. The institution can stop your payment for any reason they see fit. They may not like the product you're buying, the business you're buying from, the person you're transacting with, or the country you're sending money to.

Third-party intermediaries always have the final say. Total control over how and where you spend your money is non-existent whenever there is a third party involved in the transaction. Depending on the intermediary and situation there may even be restrictions on how and when you can withdraw your own money from your own account.

This nonsense doesn't happen with Bitcoin. *Bitcoin gives you total financial control over your money.* Other cryptocurrencies can too (maybe).

What I'm about to say is certainly not legal advice, nor am I specifically endorsing this activity. I'm just telling you what's possible. Depending on how you store your bitcoin, no person, government, or entity will *ever* be able to locate or confiscate it. You can keep that monetary value safe, secure, and hidden away for decades to come. There are a lot of ways to implement a bitcoin storage strategy. You'll learn more about this in Chapter 16.

## Self-Sovereignty

Strong cryptography, decentralization, and total financial control over one's money enables and promotes more individualism. Bitcoin is a massive step towards real *self-sovereignty.*

Self-sovereignty means that you are the maker of your own identity, online and offline, and that you're the exclusive controller of your own body, life, and data—it's autonomy, independence, self-rule. Money and finances are a major part of this equation.

Bitcoin and other cryptocurrency technologies will facilitate fully self-sovereign banking making it feasible for the first time in history. That's great for individuals, but bad for banks and governments. It's no wonder these institutions feel threatened by Bitcoin and other cryptocurrencies—your *private key* and *wallet* can replace your bank account (we'll get to what both of those mean in later chapters).

And don't forget that bankers and politicians can't meddle with cryptocurrency networks since they're decentralized and non-physical. There's no single thing or location to sanction, fine, legislate, or line up soldiers outside of.

While all this sounds great, cryptocurrencies and self-sovereign banking have a high level of personal responsibility attached to them.

Part of being your own bank means that you must look after the vault. If you lose your private keys, you'll never be able to access your bitcoin ever again. There are ways to back up your private key, but you can obviously lose the backups, too. The world of crypto is *not* like any run-of-the-mill website with a login page. You can't click "Forgot Password" if you lose or forget your private key.

Self-sovereignty is serious business and Bitcoin is only a small part of it, but damn is it exciting.

* * *

For the first time in history technology allows a paradigm shift to occur in the way we save money and transact value. What I find remarkable is that cryptocurrency and the idea of the self-sovereignty that comes along with it was predicted back in 1997 by James Dale Davidson and Lord William Rees-Mogg in their incredible book, *The Sovereign Individual.* They were way ahead of their time:

> "Now the advent of the Information Age implies another revolution in the character of money. As cybercommerce begins, it will lead inevitably to cybermoney. This new form of money will reset the odds, reducing the capacity of the world's nation-states to determine who becomes a Sovereign Individual...it will consist of encrypted sequences of multi-hundred-digit prime numbers. Unique, anonymous, and verifiable, this money will accommodate the largest transactions. It will also be

> divisible into the tiniest fraction of value. It will be tradable at a keystroke in a multi-trillion-dollar wholesale market without borders."

Some people might not want total control of their own money or have the courage to assume the responsibility needed to reach self-sovereignty. Will you aim for and work towards total financial self-sovereignty, or will you remain happily under the command of banks and financial institutions? Where you fall on this spectrum will depend on your economic, philosophical, and political views. Make your choice carefully.

Imagine living under a regime where fundamental freedom doesn't exist, where misery and hardship are the norm. In the documentary *Cryptopia*, award-winning journalist Laura Shin tells a story to the documentary's director Torsten Hoffmann about female bloggers in Afghanistan who were being paid in bitcoin. One of these women had an abusive husband. You can imagine the control he exerted over her.

She was able to secretly save enough bitcoin to the point where she was able to divorce him, get out of that chaos, and move on to a better life. If not for Bitcoin, she'd likely be stuck in that horrible situation forever. Bitcoin gave her the chance at a new life.

Or imagine living in a society that's under a constant state of surveillance and tracking, where personal privacy is not a concept. With cryptocurrency, you can begin building wealth in total privacy and prepare for bigger things. Perhaps an escape to a better country and better life is in your head. With Bitcoin, you can do it without having to carry any incriminating financial evidence on your person.

What about the millions upon millions of people across the world who aren't just in debt, but are in crushing, drowning amounts of debt? The types of debt where banks, creditors, and even scary people (like criminals, mobsters, warlords, etc.) have a claim on any new money they earn before it even reaches their pockets. Cryptocurrency offers a way to rebuild these lives in terms of finances and confidence. Sometimes having just a little bit of money makes all the difference in a person's mental state. With Bitcoin, money can be saved in such a way that creditors can never get their greasy fat fingers on it.

The total financial control and self-sovereignty offered by cryptocurrency can help turn dreams into reality. The idea of Bitcoin giving life-changing opportunity to potentially a few billion people is not some far-fetched, feel good, photo op nonsense. It's a real thing. It's powerful and it's happening.

It's easy to see how cryptocurrencies can make governments and authoritative regimes queasy. Cryptocurrencies are a major test for those in positions of power—maybe the ultimate test. Do a country's leaders really, *really* support fundamental freedoms and human rights? Or do they want control?

Watch how each country's government and leaders react to cryptocurrency over the next decade and you'll have your answer. Countries that are friendly to Bitcoin and other cryptocurrencies will see a lot of new immigration. Countries unfriendly to the technology will see massive emigration.

OK, now you're ready. You understand money, digital money, and cryptocurrency. The next two chapters discuss some basics of Bitcoin, why it's the best form of money, and lay out in broad strokes how it works.

We continue along *The Bitcoin Shortcut*. Your foundation is strong. I can see it in your eyes…things are starting to come together already.

CHAPTER 4

# BITCOIN WARM-UP

The idea of a new form of digital money is not new. Many attempts at a viable candidate were made, including DigiCash, E-gold, Hashcash, b-money, and bit gold. They all failed for various reasons.

Nothing worked until 2009 when Bitcoin launched. For the first time in history, a completely decentralized peer-to-peer digital monetary network was created. This was a difficult computer science problem that many experts didn't think was possible to solve. But the creator of Bitcoin solved it.

The network has been running smoothly since it launched, proving that decentralized finance works. *How* it works is elegant and mind-blowing. We'll get there soon enough. But first, let's look into some interesting and important aspects of Bitcoin. These will help you understand more about Bitcoin, why it was created, and why it's the ultimate form of money.

## WHO CREATED BITCOIN?

The origins of Bitcoin remain mysterious. The creator of Bitcoin posted on cypherpunk mailing lists and forums under the pseudonym Satoshi Nakamoto. A cypherpunk is defined by Wikipedia as, "any individual advocating widespread use of strong cryptography and privacy-enhancing technologies as a route to social and political change."

We don't know anything about Satoshi. Throughout the book I'll refer to Satoshi as "he" for simplicity. That's *probably* accurate though since the places Satoshi posted online were mostly filled with male computer nerds. But Satoshi could be a woman or even a team of people.

Some experts think that the sophistication of Bitcoin required a team of people to pull off. Others disagree and say that Satoshi is (or was) a lone wolf. Some potential candidates for Satoshi are alive, others are dead.

I don't want to taint your view on who could or couldn't be Satoshi since it's a fascinating topic to research on your own. It's worth going down that rabbit hole a bit and reaching your own conclusions. There's a lot of interesting information out there.

Several attempts by journalists have been made to identify Satoshi, all of which have been proven false. We'll probably never know Satoshi's identity.

But we do know that Satoshi anonymously registered *bitcoin.org* on August 18, 2008, and made his Bitcoin whitepaper available for download there on October 31, 2008. The whitepaper is a must read. You can easily find it online. Its full title is *Bitcoin: A Peer-to-Peer Electronic Cash System*.

Satoshi announced his invention on the Cryptography Mailing list on October 31, 2008, at 18:10:00 UTC with the following message:

> I've been working on a new electronic cash system that's fully peer-to-peer, with no trusted third party.
>
> The paper is available at:

> http://www.bitcoin.org/bitcoin.pdf
>
> The main properties:
> Double-spending is prevented with a peer-to-peer network.
> No mint or other trusted parties.
> Participants can be anonymous.
> New coins are made from Hashcash style proof-of-work.
> The proof-of-work for new coin generation also powers the network to prevent double-spending.

On January 3, 2009, Satoshi mined the first block of Bitcoin, creating the very first 50 bitcoin. He remained active in Bitcoin's development and launch for about two years via forum posts and some email correspondence with top cypherpunks.

His second to last forum post was on December 12, 2010, on *bitcointalk.org* when Bitcoin version 0.3.19 was released to the wild. After that, Satoshi Nakamoto vanished from the public eye.

One notable attempt by journalists to identify Satoshi was an article posted on March 6, 2014, by Newsweek titled *The Face Behind Bitcoin* which names Dorian Nakamoto, a Japanese American engineer living in California, as Satoshi. Dorian Nakamoto denies that he is Satoshi. The article led to the real Satoshi's last public correspondence.

On March 7, 2014, someone claiming to be the real Satoshi Nakamoto posted on the *p2pfoundation* forum:

> "I AM NOT DORIAN NAKAMOTO"

Some people think it was the real Satoshi, others are skeptical. From that point on, no one has heard a word from Satoshi Nakamoto.

Is Satoshi a well-known cryptographer and computer programmer hiding in plain sight? Is he an "unknown" lone genius programmer? Is Satoshi a team of people?

You must wonder, are the Japanese pseudonym, the timing of his forum posts, and the fact that he uses British English real clues to his identity? Or did Satoshi intentionally use all of these to throw people off the real trail?

As you learn more about Bitcoin, I suggest you get yourself a copy of *The Book of Satoshi* by Phil Champagne. It's a curated collection of forum posts (and some emails) from Satoshi. You get to read about technical details straight from the creator's mouth. There's a lot of value those pages. You can also read just about everything Satoshi has said in public at *https://satoshi.nakamotoinstitute.org*.

And before we continue, here's something interesting to contemplate. In an article from 2009 titled *The Education of a Libertarian*, the famed entrepreneur and venture capitalist Peter Thiel said, "the fate of our world may depend on the effort of a single person who builds or propagates the machinery of freedom that makes the world safe for capitalism."

That person is Satoshi Nakamoto.

## HOW MANY BITCOIN ARE THERE?

As of the time of this writing there are approximately 18,800,000 bitcoin in existence. Bitcoin are created during a process called *mining*, which we'll cover in full detail in Chapter 12. For now, just know that mining is the process where nodes in the Bitcoin network validate transactions and update the blockchain with new transaction blocks. Newly minted bitcoin are the reward that miners earn for doing this work.

How many bitcoin a successful miner earns depends on the current *block reward*. When Bitcoin first launched, the block reward was 50 bitcoin per block. This reward halves every four years or so, when 210,000 new blocks are mined. Currently, the block reward is 6.25 bitcoin per block.

*There will only ever be 21,000,000 bitcoin.* This supply limit is fixed in Bitcoin's code. The last bitcoin is expected to be mined around the year 2140.

So, while gold, for example, has a scarce supply that has grown every year for thousands of years and will grow for thousands of years, Bitcoin's supply is absolutely fixed. The maximum supply of 21,000,000 bitcoin is set in stone.

Never forget this: Bitcoin is the hardest money in the galaxy. Once the last bitcoin is mined in or around the year 2140 there will never be any new supply of bitcoin. That moment will mark the first time in history where any commodity or money's stock-to-flow ratio becomes infinite.

There's no need to panic about the low quantity though. One bitcoin is extremely divisible. The software allows for divisibility down to 0.00000001 BTC ($10^{-8}$ BTC). Each of these units is called a *satoshi*. Thus, there are 100,000,000 satoshis in one bitcoin.

Maximum Bitcoin Supply = 21,000,000 BTC
1 BTC = 100,000,000 satoshis
Maximum Bitcoin Supply = 2,100 trillion satoshis

Why is the supply capped at 21,000,000 bitcoin? Only Satoshi knows for certain.

One analysis suggests it's related to the M1 money supply. This is the total liquid portion of the money supply that includes cash, coins, demand deposits, traveler's checks, etc. At the time of Bitcoin's creation, the world's M1 money supply stood at $21 trillion. If Bitcoin were to totally replace the M1 supply, this would put each bitcoin's value at $1,000,000, and each satoshi at $0.01.

Another possibility is that the supply cap could be related to Bitcoin mining. The Bitcoin network regularly adjusts the difficulty of the mining process to ensure that blocks are mined about every 10 minutes on average. Considering this, about 210,000 new blocks are mined every four years, with block rewards decreasing as the total supply approaches its maximum. With the first mining cycle's block rewards set at 50 bitcoin, you could do some forward-looking math and arrive at a 21,000,000 supply limit.

Was the 21,000,000 supply limit set first or was the 50 bitcoin block reward for the first mining cycle set first? If you find Satoshi, please ask him for me.

## SATOSHI IMPROVED GOLD

Bitcoin is a store of value, a monetary network, and an asset. Up until now, the undisputed heavyweight store of value across the globe was gold. Satoshi knew this. When designing Bitcoin, he started with a model of gold and removed its defects and shortcomings. The result is the ultimate store of value: Bitcoin, a better gold.

What are the major problems with gold that Satoshi fixed? There are six.

The first problem is divisibility. Gold is poured into many forms, including multi-kilogram bricks, single-ounce bars, and even 1-gram coins (there are about 31.1 grams in 1 Troy Ounce). Dividing it all the way down to single atom units is technically possible, but certainly not practical—ultra small units would be too easy to lose, assuming you could even physically handle them at all.

As for the practicality of shopping with gold, what if you're carrying around 1-gram gold coins and need to buy something for half a gram? Or a quarter of a gram? Sure, those smaller denominated coins can be manufactured but they must be made in advance using specialized machines. Dividing gold is a hassle that requires planning and manufacturing ahead of time—gold is not instantaneously divisible. Physical gold units are not convenient.

The second problem is portability. Gold's portability is terrible. The stuff is bulky. You can't carry all your gold around town with you safely, inconspicuously, or easily. It's also very slow and expensive to safely transport gold over any meaningful physical distance.

The third problem is verifiability. Gold is not so easily verifiable. There are chains of custody involved, and special equipment and testing procedures. Not everyone has the ability to do it. If I gave you a 1-ounce bar of gold that had a stamp of authenticity on it, could you *really* tell if it was actual gold or not? How about the purity, could you confirm that? You could not unless you have the necessary equipment. And even if you have the right tools, verification errors with gold do happen.

Problem four is scarcity. There's always more gold to mine. There's enough gold in Earth's crust to keep everyone happy for a very long time. Plus, an entire industry called asteroid mining is being developed as we speak. Many nearby gold-rich asteroids have been identified. Each one of these asteroids has *much more* gold content than Earth does—orders of magnitude more. Will we eventually get our hands on it? I'd bet on it.

The scarcity of gold is overplayed and will eventually disappear. At some point in the not-so-distant future, gold may be as abundant on Earth as aluminum, which we use with throw-away mentality. This won't happen in the next 50 years. In 500, maybe. In a few thousand, you bet.

Problems five and six go together. Gold is only moderately censorship resistant and only moderately decentralized. Neither is good enough.

**Bitcoin's design solves all the problems with gold and beats all other forms of money across all metrics.** Let's look at the Money Metrics Table from Chapter 1 with Bitcoin added and discuss how Bitcoin is the best, metric by metric. Note that for each metric the higher the score, the better.

**Money Metrics Table – Bitcoin Added**

| Metric | Bitcoin | Gold | Rep. | Fiat |
|---|---|---|---|---|
| Durability | H | H | L | L |
| Divisibility | H | L | M | M |
| Portability | H | L | H | H |
| Uniformity | H | H | H | H |
| Fungibility | H | H | H | H |
| Verifiability | H | M | M | M |
| Scarcity | H | M | M | L |
| Censorship Resistance | H | M | L | L |
| Decentralization | H | M | L | L |
| Unforgeable Costliness | H | H | L | L |
| Programmability | H | n/a | n/a | n/a |

Key: H = high, M = moderate, L = low

Note: Rep stands for Representative

All the metrics in the first column of the table are qualities and characteristics that you want your money to have. The higher the score the better. As you can see, Bitcoin ranks high in every single metric. While it may look like there are some ties for some metrics, there's not. Bitcoin wins every single one. Let's walk through them.

**Durability** - While extremely difficult to do, physical gold *can* be destroyed. A targeted nuclear attack on the New York Fed would destroy over 5,600 metric tons of gold owned by 36 foreign nations and eliminate over 3% of the world's produced gold supply. There are no pinpoint attacks possible in Bitcoin. To destroy Bitcoin, the entire global telecommunications and connectivity infrastructure would have to be wiped out simultaneously and instantaneously.

**Divisibility** - As discussed earlier, each of the 21,000,000 bitcoin units is divided into 100,000,000 subunits called satoshis. This level of divisibility isn't physically possible (practically speaking) with a single dollar or an ounce of gold. Any sized transaction from the extremely small to the very large can be completed with Bitcoin instantly and without a sweat.

**Portability** - Bitcoin is a digital open monetary network with digital monetary units. Value can be transferred to anyone in the world at the speed of light. It's just as safe and easy to move one dollar worth of Bitcoin as it is to move one trillion dollars' worth of Bitcoin.

**Uniformity and Fungibility** - Bitcoin is code; it's bits and bytes. You can't get more uniform and fungible than software bits and bytes. All major forms of money in the world have high levels of uniformity and fungibility, but Bitcoin takes it to the extreme.

**Verifiability** - As you will learn throughout the book, Bitcoin's network is 100% verification and 0% trust. Its design assumes from the start that no one can or should be trusted and that *everything* must be verified by the network. There's not a monetary asset on Earth that has the robust level of verification that Bitcoin has—nothing even comes close.

**Scarcity** - The maximum supply of bitcoin is fixed at 21,000,000 total. That's far better than the limited but expandable supply of gold and the unlimited potential supply of fiat currency that can be printed at the government's will. Scarce always beats unlimited but fixed always beats scarce.

**Censorship Resistance** - Since there are no third-party intermediaries used in Bitcoin no one can tell you what you can and can't do with your money. Period. Though it's true that all in-person cash transactions can't be censored either, any financial transaction, physical or electronic, involving a third-party *is* able to be censored. Bitcoin's elimination of third parties wins this metric by a landslide.

**Decentralization** - There are tens of thousands of nodes spread all over the globe that run the Bitcoin network. None of them answer to a central authority node. No single entity, person, government, or user controls (or even influences) the operations of the Bitcoin network. This just isn't true with gold. And fiat money by definition is controlled by governments and central banks.

**Unforgeable Costliness** - In Bitcoin, there are *zero ways* to forge or spoof the cost of the work done by miners. The energy and cost required to "create" bitcoin through mining can't be faked and is easily verifiable by others in the network. Gold has a high unforgeable costliness, too. But it's just not at the level of Bitcoin's.

**Open Programmability** - Bitcoin as a network and payment system evolves because it's code. Features and improvements are made, bugs are fixed, and unnecessary or unused features are removed. It's programmable money that's optimized for its users and use cases over time. Impossible with gold and dollars.

Another highly useful feature of Bitcoin that's not listed above is that no one can steal it from you. Even if someone threatens you with guns, jail, or something else unpleasant they're not getting your bitcoin since you can take personal custody of your bitcoin *in your own head* by memorizing your private key.

With gold and fiat, they can kill you and take your money. Not true with bitcoin. Even if they do kill you, no criminal, company, bank, government, military, or terrorist can get their hands on your bitcoin if you've got the key memorized in your head. Bitcoin has a side effect of reducing violence in the world, even if just a little.

The summary conclusion:

> **Bitcoin has one job to do: to strongly and durably secure, store, and track monetary value over the long-term. It was engineered and optimized to do this job. It demonstrably does this job better than any other form of money in existence and blows gold, representative money, and fiat money out of the water.**

* * *

All of what you've learned so far was to show you the potential of Bitcoin, mainly from an economic perspective. Obviously, we haven't covered the technology yet. But with the economic intuition finished, the next chapter will prime you for the incredible details of how Bitcoin works (Part 3).

So, let's keep moving. Onwards to the final chapter of Part 2. It's a high-level overview of how Bitcoin gets the job done.

CHAPTER 5

# BITCOIN BROAD STROKES

Bitcoin is a network of computers that run the Bitcoin software. This network is responsible for validating and executing transactions for all the Bitcoin users in the world, maintaining security of the transaction ledger, and issuing new bitcoin until the maximum supply of 21,000,000 units is reached.

This fully peer-to-peer electronic cash system is the result of Satoshi Nakamoto's genius integration of ideas and innovations from cryptography and computer science. In one of his forum posts Satoshi mentions that he spent a year and half coding the project. While there are a lot of technical details in play, Bitcoin can basically be broken down into four major components:

- A decentralized peer-to-peer network
- The Blockchain
- A set of consensus rules
- A Proof-of-Work algorithm

It's the interplay of these four components that make the buying, selling, transferring, storing, and creation of bitcoin possible. Let's take a quick look at each component.

## DECENTRALIZED PEER-TO-PEER NETWORK

With Bitcoin there is no central computer server or focused point of control or operations. Instead, the network is a decentralized group of devices around the world that each run the Bitcoin protocol.

Protocols are rules and procedures for transmitting data between devices. The Bitcoin protocol is a foundational layer of code that defines Bitcoin itself and how it should function. It's open-source software that can be downloaded by anyone.

Anyone can operate as a Bitcoin miner if they want to. Miners use electricity and computing power to validate bitcoin transactions and, through consensus, immutably record them to the blockchain in the correct order. For this process to work, miners earn new bitcoin and potential transaction fees.

Some quick notes on centralized, decentralized, and distributed systems.

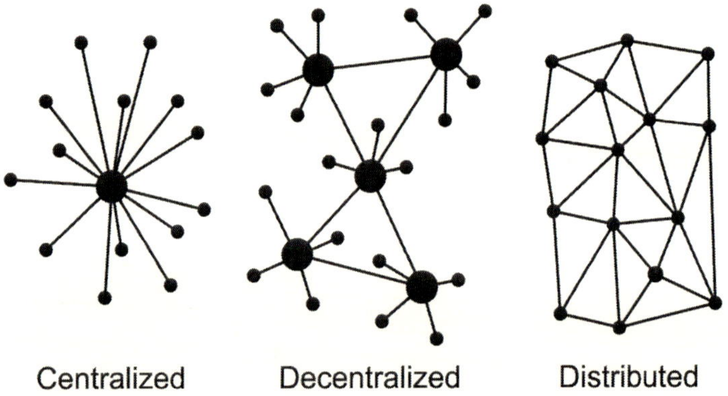


*Centralized* - All nodes connect to a single, central node or server operated by one entity. This central owner controls, stores, and provides access to the data that connecting nodes utilize.

*Decentralized* - No single entity controls the network. Data storage and computational processing are shared across multiple nodes. Each node makes its own decisions and acts independently.

*Distributed* - Generally composed of equal, interconnected nodes, but with no central owner. Computational resources and data ownership are shared equally across the network. Decision making is done in aggregate.

Centralization versus decentralization tells you where *control* of the network lies, while distributed vs non-distributed refers to the *location* of the network. For example, your bank may operate a distributed network all around the country, but the network is still centralized because only the bank controls all operations of that network.

Bitcoin is a decentralized electronic cash system that utilizes a distributed Proof-of-Work system to arrive at a consensus on the current status of its distributed public ledger, the blockchain.

## THE BLOCKCHAIN

The blockchain is a public ledger containing the entire history of Bitcoin transactions. Every full node in the network has a copy of the blockchain.

Miners group valid transactions into blocks and link these blocks together like a chain. New blocks are added to the chain roughly every ten minutes.

The structure of the blockchain and how it updates makes it easy to detect if fraudulent transactions are being attempted. Making actual changes to the existing blockchain is just about impossible to do. If a change was made to some previous block all blocks from that point forward would be invalidated and the entire chain would break.

In Chapter 11 we'll dig into many more details of Bitcoin's blockchain, how it works, and how it's so secure. We need to ease our way into that.

## CONSENSUS RULES

Bitcoin's consensus rules are precisely followed by every single node in the network to determine the validity of individual transactions and blocks of transactions. These rules allow all nodes in the network to consistently collaborate and converge upon the single correct blockchain as more and more blocks are added (as new blocks are mined multiple versions of the blockchain can exist until the correct "winning" chain emerges).

The decentralized consensus mechanism in Bitcoin is what enables every node in the network to agree on a single universal truth about the correct state of the blockchain *without any single node explicitly having to trust any other single node*. This is what allows Bitcoin to operate without the need of third-party authorities or institutions.

The details of the actual consensus rules are complicated and technical. But the general idea is simple. The rules make sure things are working as intended.

When a node verifies a single transaction, it takes that transaction through a long checklist of criteria. If any one of the criteria are not met, the node invalidates the transaction and discards it—the transaction will not propagate through the network.

Nodes also use similar rules and checklists for aggregating transactions into blocks, verifying blocks, and selecting the correct and most up-to-date version of the blockchain.

## PROOF-OF-WORK ALGORITHM

Every ten minutes on average the global network of nodes reaches consensus on the current state of transactions across the network. Arrival at this consensus is done utilizing a *Proof-of-Work* algorithm. The purpose of the Proof-of-Work algorithm is to obtain a piece of data that is costly and time-consuming to produce, yet is extremely easy, fast, and free for anyone to verify.

When you hear that someone is mining Bitcoin, it means that they are running Bitcoin's Proof-of-Work algorithm. Blocks are added to the blockchain after a miner solves a mathematical puzzle (it's not really a puzzle at all, but for now, think of it that way) and finds a solution to the block's Proof-of-Work.

Bitcoin's system of trust does not require any third parties, but it does require a lot of energy and computation. Proof-of-work provides security, prevents fraud, allows transactions to become settled (and made immutable and irreversible), and makes it impossibly expensive to alter the blockchain.

## DETAILS OF A BITCOIN TRANSACTION

A *transaction* in the Bitcoin system, also called a *payment*, is why Bitcoin exists. Transactions are the transfer of value between people in the Bitcoin network and the entirety of Bitcoin is designed so transactions can be created, validated, and permanently recorded on the public blockchain.

A Bitcoin transaction goes through four basic steps: creation, broadcasting, propagation, and validation. Let's pretend you want to send SpaceX one BTC as a deposit for your next trip to Mars. Here's what would happen.

**Creation** - You would create a transaction in your bitcoin wallet. A transaction includes three main pieces of data: the quantity of BTC to move, the bitcoin address to move the BTC to, and a digital signature. This digital signature is created with the transaction details and your private key, a secret number that only you know and have access to.

In our example, you create a digitally signed transaction asking the network to move one BTC to SpaceX's bitcoin address. Your digital signature proves that only you could have created this transaction. The digital signatures used in Bitcoin are *nothing* like handwritten signatures for paper documents. Chapter 8 covers the technical details of digital signatures.

**Broadcasting** - Once the transaction is created and signed in your bitcoin wallet it's ready to send. Once you tap the send button your wallet relays the transaction to one or more nearby nodes in the Bitcoin network.

**Propagation** - Once a node receives your transaction, it verifies it. This verification process includes a checklist of critical criteria regarding data structure, syntax, and more related rules. The node will also check its local copy of the blockchain to confirm that you have the BTC to spend and that your digital signature is valid.

If the transaction passes all the required checks, the node considers it verified and begins relaying it to other nodes in the network. Each of these receiving nodes also perform their own verification checks. Soon enough, your transaction has propagated through the entire Bitcoin network and all the nodes have independently verified it.

Verified transactions go into the Bitcoin Memory Pool (mempool). This is where all unconfirmed Bitcoin transactions sit until they're sequentially aggregated into blocks for the blockchain. Here your transaction waits for a miner to pick it up to include in the next block.

If a transaction fails verification any at point it's dropped from the network. Bad transactions die very quickly—right at the first node they reach.

**Validation** - Miners then pick up your transaction from the mempool along with many other transactions, perhaps thousands. Each miner now has a batch of new transactions including yours, but in a totally different sequential order than other miners.

In one miner's batch your transaction may be first in line but $314^{th}$ in line in another miner's batch, and $1{,}705^{th}$ in line in another miner's batch (you get the idea). This is perfectly OK.

One miner out of all of them will be able to claim that their block (and transaction ordering) is the correct one to add to the blockchain. The miner who gets to claim this is the one who wins the competition (for lack of a better word) that occurs in the Bitcoin network every ten minutes by finding a solution to the block's Proof-of-Work problem.

Once a miner finds a Proof-of-Work solution they broadcast it out to all the nodes in the network, basically saying: "My block should be added to the network and here's the Proof-of-Work." All the nodes in the network can quickly and easily verify the solution.

Once the Proof-of-Work is verified the other nodes add that block to the end of the blockchain. They immediately stop working on that group of transactions and go back to the mempool for a new batch. It's time to work on the next block in the blockchain.

* * *

SpaceX's wallet would get a notification that one BTC is being sent to it during the propagation phase. But depending on the wallet or company on the receiving end, the bitcoin can't be spent until that transaction is confirmed. When the block that contains the transaction is added to the blockchain that's a confirmation that the transaction is valid.

Let's say that the block that contains your one BTC payment to SpaceX is in block number 900,000 on the blockchain. Right after that block is confirmed miners will begin working to solve the Proof-of-Work problem for block 900,001 (which has not been added to the blockchain yet).

Once block 900,001 is validated and added to the blockchain, that counts as two confirmations for your one BTC payment to SpaceX. When block 900,002 is added, that's three confirmations for your transaction, and so on. Each confirmation adds an enormous amount of computation and unforgeable work to the blockchain. As more blocks are added, it's exponentially more difficult to reverse transactions in previous blocks. More work injected into the blockchain means it has more strength and more security.

Convention amongst the Bitcoin network is that six confirmations make a block completely irrevocable. The energy and computational power required to change a transaction six blocks deep are just unfathomable. All six blocks would have to be quickly recalculated and in such a way that the entire network wouldn't notice. It's just not happening.

Confirmations are important because they prevent double-spending. Bitcoin users protect themselves by waiting for several confirmations. Payments with zero confirmations can still be dropped from the network, so it's ill-advised to consider the payment complete with zero confirmations. If you are a merchant, you don't want to complete your end of the transaction until you see at least one confirmation.

For smaller payments one confirmation is usually considered acceptable. Once you start getting in the thousands of dollars though it's advised to wait for at least three confirmations. Top Bitcoin exchanges usually require three confirmations for a transaction to be considered final. It's only then that they allow you to then spend or trade that bitcoin. For six-figure payments it's common to require six confirmations. And for seven-figure payments and higher, it's recommended to wait for up to 60 confirmations.

A very special type of transaction called a *coinbase transaction* requires 100 confirmations. That's approximately 16 hours and 40 minutes of computational power added to the network. Coinbase transactions can only be created by miners, are the very first transaction in a new block, and are the transactions that reward miners with new bitcoin. A coinbase transactions is the only type of transaction that creates newly minted bitcoin.

There's another interesting point to keep in mind regarding Bitcoin transactions. In our example, it's not accurate to say that you are "sending SpaceX one bitcoin." It's not the same as if I walked across the room, took a dollar out of my pocket, handed it to you, and you put that dollar into your pocket.

It's common to think that a bitcoin gets taken from your wallet, sent to the network, and then sent to SpaceX's wallet, where it is stored until they use it. But Bitcoin doesn't work that way.

A point to always remember: all bitcoin are stored on the Bitcoin blockchain and are never taken off the blockchain. Only private keys are stored in bitcoin wallets, never actual bitcoin. The private keys allow you to access your bitcoin but your bitcoin never actually move off-chain.

What's actually happening in the SpaceX transaction is that you're asking the Bitcoin network to move one bitcoin from your address on the blockchain to SpaceX's address on the blockchain.

With all that said, you're now ready to start learning about the technical details of Bitcoin. The next section is how the magic happens and why Bitcoin is such a revolutionary technology. After you finish it, you'll never look at traditional banking and finance the same again. You'll start to understand what all fuss is about—the future of global financial system is beginning now.

So, buckle your seat belt. Grab a coffee. It's time to get intense. You are about to have your mind blown. Don't panic!

# Part 3

## *Technical Details of Bitcoin*

*The guts of Bitcoin. A deep dive into how and why it works. From hashing all the way to mining and everything in between. Cryptographic tools. Stretching your brain. Seeing the true genius of Satoshi's invention and having your mind blown.*

CHAPTER 6

# CRYPTOGRAPHIC HASH FUNCTIONS

All cryptocurrencies use cryptography. The word cryptography is derived from the Greek words "kryptos" and "graphein" which mean "hidden" and "to write."

When most people think of cryptography, they usually think about secret messages being passed between spies or soldiers. In this form of cryptography, an important message is *encrypted* by someone using a secret, predetermined code. This jumbled unintelligible message is then passed along to the intended recipient who then *decrypts* it. After the decryption, the recipient can clearly understand the message. This process is done to protect the contents of the message from prying eyes and adversaries.

For example, say I want to send you a secret message that says: "buy bitcoin." Beforehand, we agree that I'll use a shift cipher with a value of five. This means that for each letter in the message, I'll shift it up five places in the alphabet and use that letter instead. Here's how it works:

```
Line 1: abcdefghijklmnopqrstuvwxyz
Line 2: fghijklmnopqrstuvwxyzabcde
```

Line 1 is clearly the normal alphabet, while Line 2 is the alphabet shifted five places to the left. So, for the letter `b` I would instead write down `g` during my encryption. For `u` I would write down `z` and so on. The full encrypted message is `gzdgnyhtns`.

To decrypt the message and arrive at "buy bitcoin" you'd simply reverse the process. Anyone who knows the rule "shift the letters five places" will be able to decrypt the message and read it without a problem. An encryption like this is a two-way function, meaning that the secret message can be de-jumbled backwards to its original state. Going from encrypted to decrypted and back again as many times as you want is possible if you know the rules of the encryption.

This example uses an extremely simple encryption rule. It'd be very easy to figure out with pen and paper. For a computer, it'd be absolutely trivial to crack. But this type of cryptography, the process of encryption and decryption, is used extensively online for data security.

Of course, in the real world *way* more complicated rules and steps for encryption are used. When you connect to your online banking platform, buy from Amazon, use cloud software, or do just about anything serious online, you're using sophisticated encryption and decryption. Anytime you use https, TLS, or SSL you're using it too.

*Bitcoin does not use encryption and decryption.* It doesn't need to. This may sound strange but in Bitcoin nothing needs to be encrypted. Bitcoin uses other tools of cryptography to protect the monetary funds stored on the network.

The first one is the *cryptographic hash function.* This tool is the perfect starting point for our technical discussion of Bitcoin because cryptographic hash functions are used *a lot* in Bitcoin. We can't meaningfully progress through Bitcoin's technology until you have a solid understanding of this tool first.

## WHAT A CRYPTOGRAPHIC HASH FUNCTION DOES

A cryptographic hash function is an algorithm that takes a digital input of any size and transforms it into a fixed-length string of numbers and letters that appear random. The output of this process is called a *hash*. The hash can be considered the digital input's fingerprint.

If you pressed the tip of your finger onto a fingerprint scanner it would produce the same exact fingerprint every single time. You could be identified by that fingerprint. It's also true that no two people in the world have the same fingerprint. The probability of that happening is so small it's considered zero.

The same is true with digital inputs and cryptographic hash functions. Whatever the particular input is, it will always produce the same hash over and over. And no two known inputs will ever produce the same hash.

The size of the input is also inconsequential. It could be a single letter, a sentence, a five-page document, or even a billion-page document—the output hash will always be the same length.

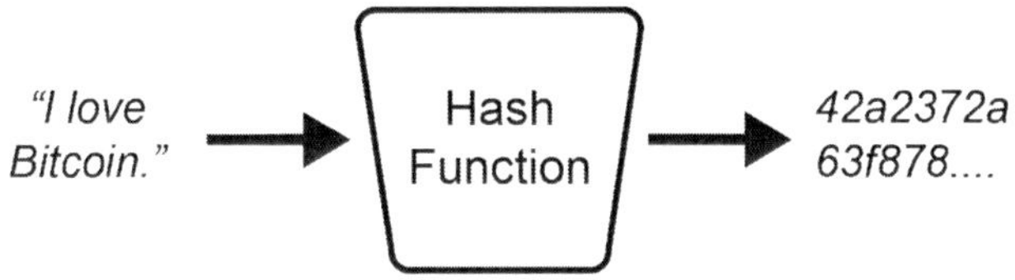


The hash output is represented as a *hexadecimal* number. The hexadecimal system is a base 16 system which means it uses sixteen different symbols to represent its sixteen values. In hexadecimal, values `0` through `9` represent the numbers 0 to 9, and letters `a` through `f` represent numbers 10 through 15.

The range is `0-f`, representing 0-15, has sixteen different values total.

Here are two examples of hexadecimal converted to regular decimal numbers. The value `3c` equals 60, and the value `3c7f` equals 15,487. Let's look at how you arrive at the decimal number from the hexadecimal. The conversion from hexadecimal to decimal uses three things: the hex number, its place value in the number, and the base value 16.

In the `3c` example below, `c` equals 12 in decimal. This 12 is multiplied by 16 raised to the zeroth power (most of the computer world starts counting at zero, not one) since its place value is one. The next hexadecimal number, `3`, represents 3 in decimal. Its place value is two, so it is multiplied by 16 to the first power. The second example shows this logic extended. Hexadecimal is not as difficult as it first seems.

`3c` $= (3 \times 16^1) + (12 \times 16^0)$
$= 48 + 12$
$= 60$

`3c7f` $= (3 \times 16^3) + (12 \times 16^2) + (7 \times 16^1) + (15 \times 16^0)$
$= 12{,}288 + 3{,}072 + 112 + 15$
$= 15{,}487$

Why is hexadecimal used at all? To answer that let's start with the *bit*. A bit is the smallest amount of usable information in a computer. A bit's value is either 0 or 1. When we group 8 bits together, we get a *byte*. For example, `0000 0000` is a byte. Also, `0000 0001` is a byte, or `1111 1111`. Often a space is put in between groups of 4 bits to make it easier to read.

Every combination of 0's and 1's ranging from `0000 0000` up to `1111 1111` is possible. There are 256 potential values of a byte, ranging from 0 to 255. The first four bits have sixteen different combinations, and the second four bits have sixteen different combinations.

That's 16 x 16 = 256. This is where the base 16 hexadecimal system comes in handy for computers. Each byte is represented as two hex digits.

| Binary | Hex | Decimal | Calculation |
|---|---|---|---|
| 0000 0000 | 00 | 0 | $(0 \times 16^1) + (0 \times 16^0)$ |
| 0000 0001 | 01 | 1 | $(0 \times 16^1) + (1 \times 16^0)$ |
| ... | ... | ... | ... |
| 1111 1110 | fe | 254 | $(15 \times 16^1) + (14 \times 16^0)$ |
| 1111 1111 | ff | 255 | $(15 \times 16^1) + (15 \times 16^0)$ |

The most used cryptographic hash function in Bitcoin is called Secure Hash Algorithm 256-bit (SHA256). The output of SHA256 is a 256-bit number, which is 32 bytes long (256 bits divided by 8 bits equals 32 bytes). Thus, it outputs a number that is 64 hex digits long. In other words, SHA256 outputs a number between zero and an unbelievably gargantuan value. How gargantuan? The number of SHA256 outputs is:

$$2^{256} = \begin{array}{l} 115{,}792{,}089{,}237{,}316{,}195, \\ 423{,}570{,}985{,}008{,}687{,}907, \\ 853{,}269{,}984{,}665{,}640{,}564, \\ 039{,}457{,}584{,}007{,}913{,}129, \\ 639{,}936 \end{array}$$

That's roughly the same number of atoms in the observable universe. It's a lot of possible numbers.

Let's look at some example inputs and output hashes of the SHA256 hash function:

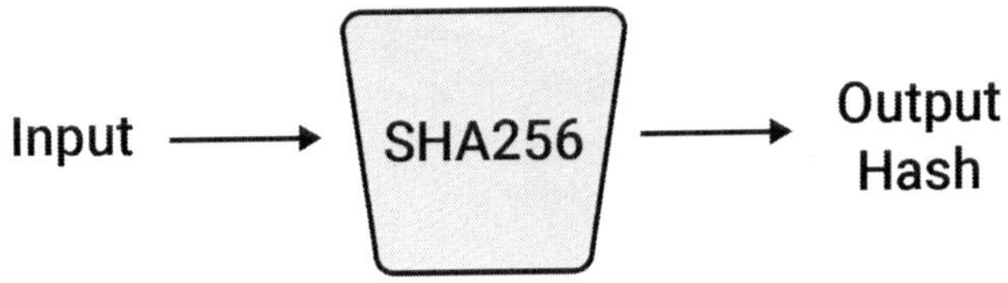


| | |
|---|---|
| **Input:** | Hello |
| **Output Hash:** | `185f8db32271fe25f561a6fc938b2e26`<br>`4306ec304eda518007d1764826381969` |

| | |
|---|---|
| **Input:** | He ate 4 apples. |
| **Output Hash:** | `8cf5f1934ffc5b160f172795aea35474`<br>`3d88757509252507ddcffe06a303659` |

| | |
|---|---|
| **Input:** | He ate 5 apples. |
| **Output Hash:** | `bb6eabbc31523128ee9527cda2d455ef`<br>`80f13871fd834c67bbbc6eb276f5644b` |

| | |
|---|---|
| **Input:** | The Bitcoin Shortcut is the best book on Bitcoin ever written. Everyone should read it as soon as possible, order copies for all of their friends, and give it a 5-star rating on Amazon. Tell everyone you know about The Bitcoin Shortcut! |
| **Output Hash:** | `a37af89acee70db4cf77aaf296144d60`<br>`da737080eefa897c2cb212f8f21a9d1e` |

Notice how all the hashes are the exact same length, regardless of the size of the input. This will always be the case.

Also, look at inputs two and three. Just by changing one character, the resulting hashes are completely different.

**Cryptographic hash functions are one-way functions.** Generating the hash from the input is quick and easy but going backwards, starting with the hash, and calculating the input, is impossible. There's absolutely no way at all to reverse engineer what the input is based on the hash.

For a more detailed look into how SHA256 creates hashes please see the Appendix.

## PROPERTIES OF A HASH FUNCTION

A hash function takes an input (also called a pre-image), performs a series of steps on that input, and outputs a hash (also called a digest). For a hash function to be considered usable in the real world it must have the following five properties:

1. The input can be any size.
2. The output is always the same fixed size.
3. The same input always produces the same output.
4. A slightly different input produces a very different output.
5. It's efficiently computable; the output can be calculated in a reasonable amount of time.

## SPECIAL PROPERTIES OF CRYPTOGRAPHIC HASH FUNCTIONS

Cryptographic hash functions are a special class of hash functions. For a hash function to be cryptographically secure, like SHA256, it must have three additional properties:

- Collision Resistance
- Pre-image Resistance (also called Hiding)
- Second Pre-image Resistance (also called Puzzle Friendliness)

**Collision Resistance**

A collision occurs when two unique inputs produce the same exact hashes. Collision resistance means that nobody can find two unique inputs that result in the same output. More accurately, collision resistance means that it's *extremely* difficult to find a collision.

Totally preventing collisions is not possible. Every hash function has collisions. Take our beloved SHA256 hash function that Bitcoin utilizes. There are $2^{256}$ unique outputs for SHA256. If we selected $2^{256} + 1$ unique inputs, we are guaranteed to find at least one collision. This is true because the number of inputs exceeds the number of possible outputs.

For a hash function with $2^{256}$ outputs, you'd be certain to find a collision trying out $2^{256} + 1$ inputs and you'd expect to find one, on average, by trying $2^{128}$ inputs.

But, and there's a BIG but, this would take an unfathomable amount of time. To put that amount of time into perspective consider that the age of universe is approximately 14 billion years. That's $1.4 \times 10^7$ years. If you had a computer calculating 100,000,000 hashes per second it would take approximately $10^{23}$ years on average to find a collision of SHA256. That's 7,142,857,142,857 *times* as long as the age of the universe. All the computers in the world *combined* running since the beginning of the universe would still have such an extremely low probability of finding a collision that it's effectively zero.

You don't have to worry about a collision with SHA256.

### Pre-Image Resistance (Hiding)

Recall that *pre-image* is another term for the hash function's input. Pre-image resistance means that if you know the hash function used and you have the output hash it's infeasible to find the input that results in that hash.

Pre-image resistance also implies that there is no *likely* input that results in a particular output and that the only optimal way to find the input for a given hash is to try every possible input randomly.

### Second Pre-Image Resistance (Puzzle Friendliness)

This is the most complicated of the three special properties. Second pre-image resistance means that if you have a pre-image, the hash function, and the hash, it's very difficult to find *another* input value that produces the same hash. A hash function that is puzzle friendly will not show any pre-determinable correlation between a pre-image and a hash. In other words, you can't select a certain input with hopes of returning a specific hash.

Second pre-image resistance and collision resistance are similar but have an important difference. Let's look at both using inputs $x_1$ and $x_2$. With collision resistance, an attacker on the network could find *any* two inputs out of the entire input set that meet the requirement: hash of $x_1$ equals hash of $x_2$.

With second pre-image resistance, the attacker is *given* $x_1$ and must find an $x_2$ such that hash of $x_1$ equals hash of $x_2$, utilizing some sort of exploitable relationship between the input, the function, and the hash.

The following figure illustrates the properties of collision resistance, pre-image resistance, and second pre-image resistance.

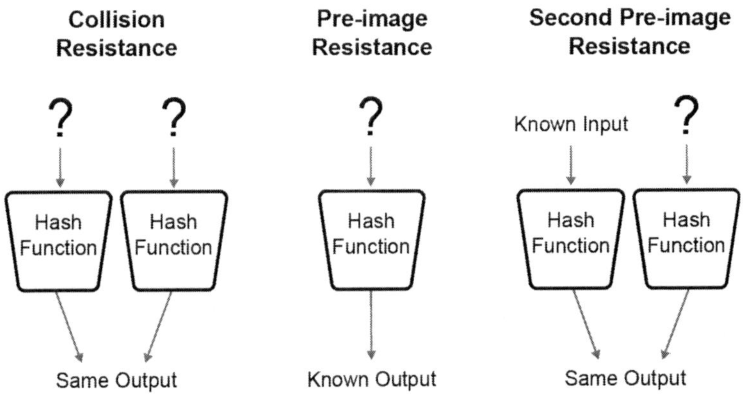


The properties of cryptographic hash functions sum up to one critical, core idea: **the only way to find an input that produces a specific hash is through brute force trial and error.** You must try *every single input.*

These properties and the security they provide are very important to Bitcoin since cryptographic hash functions are used within the network to verify that data has not been tampered with or changed. I've said that Bitcoin uses SHA256, but the truth is that what's mostly used in Bitcoin is double SHA256.

This means that an input is first hashed using SHA256 and then *that* hash is input into SHA256 a second time. While we don't know for certain why Satoshi chose to use double SHA256 the most agreed upon idea was that he likely used it to prevent something called a *length extension attack.* The details of a length extension attack are very interesting, but we won't be covering them since they're beyond the scope of Bitcoin and this book.

* * *

Let's summarize. A cryptographic hash function is a one-way function that takes an input of any size and outputs a fixed-sized string of numbers and letters that appear random. The hash is equivalent to a digital fingerprint and it's impossible to determine the input based on this digital fingerprint.

Cryptographic hash functions are very secure due to a variety of properties and the only known way to find the input that results in a particular hash is through brute force trial and error. Bitcoin uses cryptographic hash functions in many places.

Congrats! With your new understanding of this cryptography tool, you're ready to continue tackling the technology behind Bitcoin.

# Pause

*In Bitcoin, ownership of individual bitcoin units is established and proven using three distinct things: digital keys (private and public key pairs), digital signatures, and bitcoin addresses. Let's dive into each one, starting with digital keys.*

## CHAPTER 7

# DIGITAL KEYS

In Bitcoin, digital keys are part of the mechanism that enable monetary value to be moved around the network from person to person. Digital keys always come in pairs consisting of a private key and a public key.

A private key is a large secret number that allows bitcoin to be spent while a public key is a large non-secret number that allows bitcoin to be received. Bitcoin transactions are signed with the help of private keys and verified with the help of public keys.

It's been said that a public key is like a bank account number and a private key is like the PIN to that account. That's a pretty good analogy for helping wrap your head around the idea. But keep in mind that each key is a 256-bit number. They're *far more secure* than a simple bank account number and frail little 4-digit PIN.

Digital keys are typically created by a user's wallet software and stored within that wallet. They are not part of Bitcoin's blockchain.

Recall that to spend bitcoin, a transaction must be created that includes the quantity of bitcoin to move, the address to move the bitcoin to, and a digital signature.

Here's how digital keys play a role in a transaction:

1. The private key is used to generate a digital signature.
2. The public key and digital signature are presented in the transaction.
3. Using the public key and digital signature, the network can verify the transaction, confirming that the person who created the transaction has the bitcoin available to spend.

A private key is a *critical* number to *you.* Possession of a private key gives control over the bitcoin associated with that key.

There's a phrase in the Bitcoin community popularized by Andreas Antonopoulos, the author of the technical book *Mastering Bitcoin*, that goes like this: "Your keys, your bitcoin. Not your keys, not your bitcoin."

What he means is that without owning and controlling your keys, you don't actually own and control your bitcoin. There are many services out there that allow you to buy and sell bitcoin, but *don't* provide you with the private keys. This means that you don't *really* own the bitcoin—it's held and controlled by them. It's basically just an IOU in fiat from the company to you.

Not your keys, not your bitcoin. Please say it out loud: not your keys, not your bitcoin. Never forget this.

With that said, there are two very important things to keep in mind regarding your private keys. They're so important, they need to be emphasized.

> **Your private keys must be kept secret, no matter what. Always stay vigilant. If someone else gets your private keys, they control your bitcoin and can transfer it at their discretion.**

and:

**It's your responsibility to back up your private keys. If you lose your private keys, the bitcoin secured by those keys are lost forever. In Bitcoin there is no equivalent of a "reset password" button for private keys.**

We'll discuss a lot of details about backing up private keys later in the book. You'll learn exactly what to do. It's easy enough but it's a great reminder that with Bitcoin comes a new level of personal responsibility. The more responsible you are, the more financial self-sovereignty you can attain.

## HOW PRIVATE AND PUBLIC KEYS ARE CREATED

A private key and public key have a mathematical relationship that allow them to play the roles in transactions that they do. The private key is a random number. From the private key, the public key is generated. Then, from the public key, a bitcoin address is generated.

The following figure shows this relationship. We'll cover exactly what happens in each of these steps in this chapter and in Chapter 9 - Bitcoin Addresses.

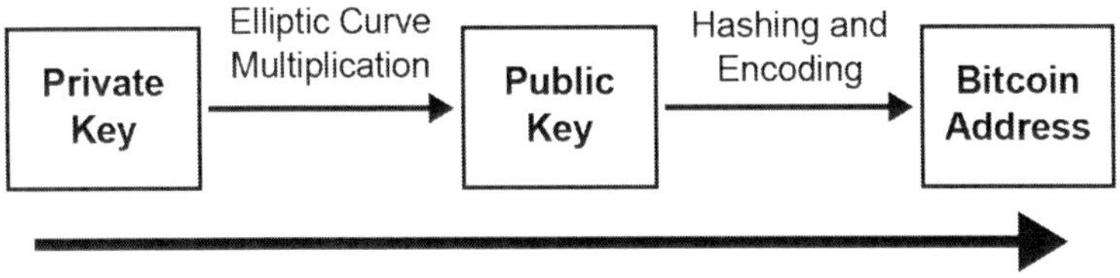


This is a one-way, forward process. It's impossible to go backwards.

It's worth noting that the processes of generating the public key and bitcoin address are one-way only. It's impossible to compute the public key if given the bitcoin address. It's also impossible to compute the private key if given the public key or bitcoin address.

### Private Key Generation

A private key is a 256-bit number created by a random number generator. Bitcoin wallets create them for you, but you could use a random number generator on your computer or online to generate your own 256-bit number if you want. Just be sure the generator has enough *entropy.* See the Appendix for more on entropy and security strength.

If you have spare time, you could even flip a coin 256 times and record the outcomes. Each "heads" counts as a 0 and each "tails" counts as a 1. Here's an example private key I generated:

```
11110001010110011111010100110
00110101011000110010100001010
00010111111000010001110110010
00110011010110110101010010001
00011011001111000101101101000
01101000000110110010010001111
01000111100000011000100111000
01111010010000011010110100010
00100111101110111011101
```

Intuitively, you probably already guessed that the private key is a random number between 0 and $2^{256}$ - 1. But it's not quite that high. Instead, a private key can be any number between 0 and n - 1, where n is 1.1578 x $10^{77}$. This number is less than $2^{256}$, but still enormous. The private key needs to be lower than this number because it's the "order of the elliptic curve" that Bitcoin uses to generate public keys.

So, when a private key is generated, we check to see if it's less than this upper limit. If it is, we keep it and have a suitable private key. If not, we just generate a new private key and keep doing so until we find one that's within the acceptable range.

For simplicity, the private key is often represented in hexadecimal form. Our example private key from above converted to hexadecimal is:

```
78acfa98d58ca142fc23b2335b5488d9
e2da1a06c91e8f031387a41ad113dddd
```

That's all a private key is, just a random number selected from a gigantic set of possible numbers. And if you're wondering, "what if two people have the same private key?", you can stop thinking about it. There's effectively a zero percent chance that any two people on Earth have or ever will have the same private key. But more on this later in the book.

### Public Key Generation

A *public key* is generated from the private key using a cryptographic algorithm called *Elliptic Curve Digital Signature Algorithm* or ECDSA for short. The private key goes through an elliptic curve multiplication using a specific curve called *secp256k1.*

We won't cover the full details of how this works. It's too complicated. We'll just discuss some broad strokes. Try not to be overwhelmed; you can handle this.

Here are a few important points regarding public key derivation using ECDSA:

- It is a one-way function, just like cryptographic hash functions.

- A unique private key always produces the same public key.
- The elliptical curve multiplication is, $K = k * G$, where $k$ is the private key, $G$ is the generator point, and $K$ is the public key.
- Calculating $k$ if you know $K$ can only be done by brute force; you must try all values of $k$.
- Since computing the private key from the public key is infeasible, public keys can be shared freely with anyone and made public, hence the name.

Elliptic curve cryptography utilizes addition and multiplication on an elliptic curve. For *secp256k1*, the exact curve used in Bitcoin, a public key is a whole-number solution to the following equation:

$$y^2 = x^3 + 7 \bmod p$$

$$\text{where } p = 2^{265} - 2^{32} - 2^9 - 2^8 - 2^7 - 2^6 - 2^4 - 1$$

There are roughly $2^{256}$ solutions, meaning there are about that many public keys possible. While we won't walk through a full example of the calculation, as it would be distracting, here are the general steps involved in calculating a public key.

On the following page is an example image of an elliptic curve like the one used in Bitcoin. You can think of the *secp256k1* curve as a similar but more complex curve, on an unimaginably large grid.

Recall the equation $K = k * G$. We already have our private key, $k$, generated. Next, we multiply it by $G$, which is known as the generator point. The generator point for *secp256k1* is a predetermined point on the elliptic curve that has x and y coordinates. It's a standard point used for every Bitcoin public key calculation. Keep in mind that it's still impossible to calculate backwards and arrive at the private key if you have the public key, even though *G* is standard.

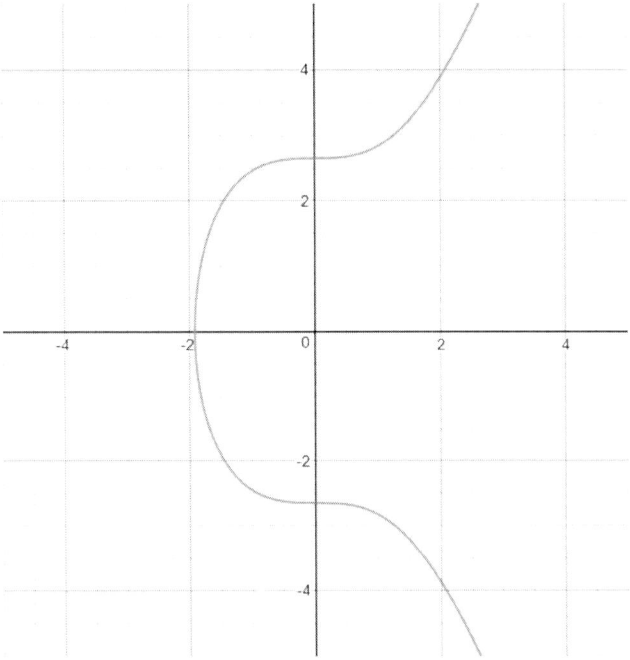


The generator point for Bitcoin has the following coordinates:

x = 55066263022277343669578718895168534326250603453777594175500187360389116729240

y = 32670510020758816978083085130507043184471273380659243275938904335757337482424

In essence, to perform the elliptic curve multiplication we use addition and doubling, which is defined as adding a point on the curve to itself. The calculation involves line intersections, tangent lines, reflecting points across the x-axis, and so on—too much to get into. But once the steps are completed there's no obvious relationship between the private key and the generator point. The public key has been safely calculated one-way.

We start with a large random number (your private key) and multiply it by a starting point on an elliptic curve (generator point). *The result of this multiplication, a new (x, y) point on the curve, is your public key.*

Continuing with our example, we now have the following private and public keys:

Private Key:

```
78acfa98d58ca142fc23b2335b5488d9
e2da1a06c91e8f031387a41ad113dddd
```

Public Key:

```
028aa24fdc478d047af3311bc576c28a
940c58d59ae9487f3ea67c056ff40c33
7e
```

There are two curious things to note about the public key. First, it sure doesn't look like an (x, y) coordinate, does it? It's a hexadecimal number just like the private key. What gives? Second, while it's a little tough to notice, the public key is 33 bytes long while the private key is 32 bytes long. Where's that extra byte coming from?

Because of the symmetry of the elliptical curve there are precisely two points on the curve for each *x*-value. One point has an even number *y*-value and the other has an odd number *y*-value. The point of this (no pun intended) is that it's not necessary to worry about the exact *y*-value, only whether it's even or odd. The *x*-value is what really matters.

If the *y*-value is even, we take note and write down a 02. If it's odd, we use 03. This even/odd "note" is called a prefix. We don't need the whole (x, y) coordinate, just the *x*-value and the prefix. The public key is the hexadecimal of the *x*-value with the prefix attached to the very beginning of it. See how our public key starts with 02? This prefix is the extra byte. And since we only need the x-value and the prefix, we can convert this new single number to hexadecimal.

See that wasn't too bad, was it? A private key is a big random number, and the public key is another big number generated from that private key. As I mentioned earlier in this chapter, the public key is then used to generate a bitcoin address. We'll cover bitcoin addresses two chapters from now. But first you'll learn about digital signatures. They're a tool used to prove to the Bitcoin network that you approve a transaction and want to move some bitcoin.

One last thing before we move on to digital signatures. I just wanted to say:

> **Your keys, your bitcoin. Not your keys, not your bitcoin. Always keep those private keys backed up, safe, and secure!**

CHAPTER 8

# DIGITAL SIGNATURES

Digital signatures are used in Bitcoin to prove that you confirm and approve that a transaction using your bitcoin should occur. When you first hear the words *digital signature* you might imagine typing your name into a text field on a website, pasting an image of your handwritten signature to a PDF document, or even scribbling your name with your finger on a smartphone or tablet.

It's a very important distinction to note that those are examples of electronic signatures *not* digital signatures. While digital signatures are electronic in form, they are vastly superior to all other types of electronic signatures because they are mathematically linked to some content, called a message, and a private key.

*Digital signatures prove that someone owns a private key without revealing that private key to anyone.*

## PROPERTIES OF DIGITAL SIGNATURES

Digital signatures are required to have two properties to be useful: verifiability and unforgeability.

**Verifiability**

The Bitcoin network must make sure it's *you* who is signing the message. In the physical world, important handwritten signatures require a notary public to witness a signature and verify the identity of the signer through photo identification. In the Bitcoin network anyone can easily and independently verify that it's you who created a digital signature with the help of your public key.

So, while only you can *create* your digital signature using your private key, anyone in the Bitcoin network can *verify* your digital signature using your public key. The person verifying your signature will never see your private key and doesn't even need to know it's you who created the digital signature. The verification only proves and validates that the digital signature in question was made using the private key it claims to be made with.

No third parties are needed in the creation or verification of a Bitcoin digital signature.

**Unforgeability**

Physical handwritten signatures are *meant* to look the same on all documents since the signature specimen is supposed to be directly linked to a physical person. If I look at a variety of documents signed by someone, I'm supposed to be able to look at all those scribbles and tell that those signatures were created by the same person.

The problem with physical handwritten signatures is that they don't offer as much protection as people think they do. It doesn't take much practice to forge a scribble. It's also incredibly easy to cut a signature from one document, tape it to another, and make a photocopy. This new "signed" document may be a document the creator of the signature has never seen and doesn't endorse.

Poor security troubles traditional electronic signatures, too. Electronic scribbles are easy to duplicate, pre-created images of signatures are easy to copy and paste anywhere and typing a name into a text field is as simple as spelling the name correctly.

Third parties can help alleviate these issues for physical and electronic signatures, but they can't protect against all types of forgery.

In Bitcoin, digital signatures are much more sophisticated. They are connected to specific transactions and can only be used once. This is because the signature for a transaction is created using your private key and the actual data inside the transaction. A valid signature for a specific transaction *cannot* be copied and used to indicate you endorse a different transaction—the network would sniff that out faster than you can say Bitcoin.

Every transaction you initiate will have its own unique digital signature. And since we're dealing with one-way cryptographically secure processes, if a thief or adversary has your public key and has seen one of your previously verified signatures, they won't be able to figure out your private key and use it to sign a fraudulent transaction in the future.

Bitcoin digital signatures are unique, single-use, transaction specific, and are computationally infeasible to forge.

## PURPOSES OF DIGITAL SIGNATURES IN BITCOIN

Digital signatures are used in Bitcoin for three purposes: authorization, undeniability, and integrity.

First, they prove that the owner of the private key associated with specific bitcoin units wants to spend those bitcoin units—they *authorize* the transaction.

Second, this proof of authorization can't be denied. In other words, the creator of the transaction can't dispute that he or she created it, saying "oh, I didn't make this payment!" This *undeniability* is also called non-repudiation.

Third, the digital signature ensures *integrity*. After the transaction is signed no part of that transaction can be modified by anyone without the transaction being invalidated, not executed, and thrown out.

## DIGITAL SIGNATURE CREATION AND VERIFICATION

At the time of this writing the Elliptic Curve Digital Signature Algorithm (ECDSA) is used for creating and verifying digital signatures in Bitcoin. Sometime around late 2021 or early 2022 new protocol changes will be implemented in Bitcoin which include the *Schnorr* signature scheme. While an incremental improvement, Schnorr digital signatures will be more advantageous on a technical level, increase the level of privacy in Bitcoin, and make the network a bit more scalable.

Given this change, I'll only briefly cover the current way that digital signatures are currently created and verified. From a user perspective, you won't notice any difference between ECDSA and Schnorr.

Schnorr and ECDSA use the same elliptic curve, *secs256k1*, and compatibility with ECDSA must be kept going forward, otherwise old nodes on the network would consider Schnorr signatures invalid and new nodes would consider all previous transactions using ECDSA invalid. What you are about to learn will not soon be obsolete and both methods use the same exact private and public keys.

## Digital Signature Creation

A digital signature is created with a signing function that uses information from the transaction and your private key. First, a subset of the information from the transaction is hashed using double-SHA256. Next, using this hash and a signing key (your private key is used), the signing algorithm performs some complicated mathematics and spits out two 256-bit values, commonly referred to as *r* and *s*. This is your digital signature.

$$\text{Digital Signature} = (r, s)$$

Here's an example of the signing function output. Note the `0x` prefix is used in code to indicate that the following number is written in hexadecimal:

*r* = `0x37206a0610995c58074999cb9767b87`
`af4c4978db68c06e8e6e81d282047a7c6`

*s* = `0x8ca63759c1157ebeaec0d03cecca119`
`fc9a75bf8e6d0fa65c841c8e2738cdaec`

These two pieces, *r* and *s*, are encoded using a scheme called Distinguished Encoding Rules (DER), which is an international standard. Using *r* and *s* from above, the encoded signature becomes:

```
304502203720 6a0610995c58074999cb9767
b87af4c4978db68c06e8e6e81d282047a7c6
0221008ca63759c1157ebeaec0d03cecca11
9fc9a75bf8e6d0fa65c841c8e2738cdaec
```

Recall that a Bitcoin transaction contains the quantity of BTC to move, the bitcoin address to move it to, and a digital signature. We haven't covered bitcoin addresses yet, so while transactions don't really look like this, let's construct a fake example transaction.

We'll come back to it in just a second:

*Message*: "Please send 5 BTC to Craig. Signed, Sarah."

*Signature*:
```
3045022037206a0610995c5807499
9cb9767b87af4c4978db68c06e8e6
e81d282047a7c60221008ca63759c
1157ebeaec0d03cecca119fc9a75b
f8e6d0fa65c841c8e2738cdaec
```

## Digital Signature Verification

Anyone can verify the digital signature attached to a transaction. To do it, they need the signature, the transaction data, and the public key. Note that the public key must be the one that corresponds to the private key which was used to create the digital signature that's being verified. If the wrong public key is provided the verification will fail.

Using these three pieces of information the ECDSA signature verification algorithm will tell the verifier if the public key and digital signature were made using the same private key. If the verification algorithm doesn't fail, we know for certain that the person who owns the private key that generated the provided public key is the same person that created this transaction's digital signature. If verification succeeds, we know for sure that the person who owns these bitcoin authorizes this transaction.

Just as a reminder, there is no encryption used in Bitcoin. It is not needed. ECDSA does not use encryption in its signature algorithm. The creation and verification processes use elliptic curve mathematics to achieve their objectives. While the private key is used during the creation of a digital signature, it is not known, used, or revealed during the verification process.

Yet, digital signature verification can confirm that the person who owns the private key that generated *this* public key is the same person who generated *this* digital signature without ever knowing the person's identity or private key. Pretty damn slick.

## A FRAUDSTER FAILS

Let's take another look at the transaction example where Sarah sends 5 bitcoin to me.

Message: "Please send 5 BTC to Craig. Signed, Sarah."

Signature:
```
3045022037206a0610995c5807499
9cb9767b87af4c4978db68c06e8e6
e81d282047a7c60221008ca63759c
1157ebeaec0d03cecca119fc9a75b
f8e6d0fa65c841c8e2738cdaec
```

Imagine that somehow a fraudster named Leonard intercepts this transaction after it was sent to the network and was somehow able to change it, so he gets paid instead of me.

It would then look like this:

*Message*: "Please send 5 BTC to Leonard. Signed, Sarah."

*Signature*: `304502203720 6a0610995c5807499 9cb9767b87af4c4978db68c06e8e6 e81d282047a7c60221008ca63759c 1157ebeaec0d03cecca119fc9a75b f8e6d0fa65c841c8e2738cdaec`

Since the signature is created using the message content, the `304502`...`8cdaec` signature is no longer valid. This altered transaction would instantly be thrown out by the very first node it reached in the network. Sarah's 5 bitcoin would be safe, and Leonard would be left scratching his head and contemplating his life choices.

It's worth reiterating that even changing a single character in a message results in a *totally* different signature. Imagine that Leonard and I are working as team. He intercepts Sarah's transaction and changes it so I received one extra bitcoin that I can share with him:

*Message*: "Please send 6 BTC to Craig. Signed, Sarah."

*Signature*: `3045022037206a0610995c5807499 9cb9767b87af4c4978db68c06e8e6 e81d282047a7c60221008ca63759c 1157ebeaec0d03cecca119fc9a75b f8e6d0fa65c841c8e2738cdaec`

This altered transaction would also be immediately discarded by the network since the `304502`...`8cdaec` signature is not valid either. You have to love the integrity Bitcoin digital signatures provide.

* * *

Since you've got cryptographic hash functions, digital keys, and digital signatures under your belt it's time to learn about bitcoin addresses. In practice, your transaction would never say something like "Please send 5 BTC to Craig." Instead, you'd construct the transaction so it sends the bitcoin to my bitcoin address.

Bitcoin addresses are quite clever and one of my favorite parts of Bitcoin. I bet you'll like them, too.

CHAPTER 9

# BITCOIN ADDRESSES

A *bitcoin address* is where people send you bitcoin. Your address can be shared publicly with anyone you want without worry that anything secret is being revealed. It's a relatively short string of numbers and letters that is generated from your public key.

Using our sample public key from Chapter 7, the associated address is:

```
18yKJg4MXW2ByhjHMpj1iZPpAuaMidgd4v
```

You'll see exactly how this jumble of characters is generated but first let's quickly talk about address formats. Currently, there are three different formats used in Bitcoin. Every available bitcoin wallet and cryptocurrency exchange supports at least one of these address formats:

- P2PKH - Bitcoin's original address format. Stands for Pay-to-PubKey Hash. Always begins with a `1`.

- P2SH - Stands for Pay-to-Script Hash. Commonly used for multisig addresses. Always begins with a `3`.

- Bech32 - Recent address format released in 2017. The native segwit address format. Always begins with `bc1`.

Reading up on the pros and cons of these address formats can be a useful exercise that will help you understand Bitcoin even more.

As you can see, our bitcoin address starts with a `1`, so it's a P2PKH address. All the details you're about to read in this chapter apply to P2PKH addresses.

## HOW AN ADDRESS IS GENERATED

Bitcoin addresses are extremely cool. An address is data hashed and encoded in such a way that it protects against typing errors, avoids ambiguity, and makes it easier for humans to read. There's a lot to unpack.

Once you have your public key (don't forget that it's generated from your private key), it goes through the following three steps:

1. Hash the public keys using SHA256

2. Take the hash from step 1 and hash it with RIPEMD160, resulting in your Public Key Hash (PKH)

3. Perform Base58Check encoding on the PKH

The following page shows a quick visual of how a bitcoin address is generated.

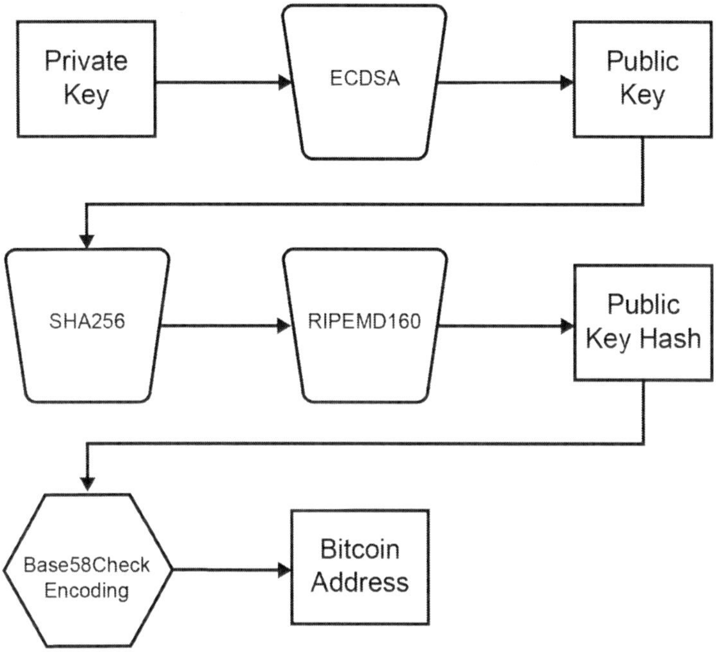


This process takes a 33-byte public key and shortens it while maintaining security and making the address easier for humans to use.

Let's walk through the full example. Recall that our example 33-byte (264 bit) public key, the starting point, is:

```
028aa24fdc478d047af3311bc576c28a9
40c58d59ae9487f3ea67c056ff40c337e
```

First, we go ahead and hash this public key using SHA256, which you're already familiar with. As a reminder, SHA256 outputs a 32-byte (256 bits) hexadecimal number.

The SHA256 hash of our public key is:

```
3994c91de8f4c8f28df654a884efe58b
8562ab334e9066a3b3a8408ae64bbc0a
```

Next, we take this result and hash it using the RACE Integrity Primitives Evaluation Message Digest (RIPEMD) 160 cryptographic hash algorithm. This is also a one-way cryptographic hash function. It outputs a 20-byte (160 bit) hash. The result of this second step is known as the *public key hash* (PKH).

In our example the PKH is:

```
576f74506ff5e575fb1c
4b42a145a8502977bee6
```

At this point, the public key has gone through a double cryptographic hash. Why use the combination of SHA256 and RIPEMD160? The truth is, we don't know. Satoshi never explicitly stated why he did this. Bitcoin addresses are the only place in Bitcoin where RIPEMD160 is used. Everywhere else it's SHA256.

This shorter byte size certainly helps with scaling though. Every byte matters when you are dealing with a global electronic monetary network that has the long-term potential to be the base layer store of value for 8 billion people.

Security remains strong. Even if an attacker breaks RIPEMD160, he still must also crack SHA256 and vice-versa.

It's worth mentioning that a single collision for either of these hash functions has never been found. This combination of security and size reduction is likely what Satoshi had in mind.

But at this point a problem for remains for the users of Bitcoin. Imagine that you wanted to send 5 bitcoin to me using the PKH we have generated but on the fourth character you accidentally typed an `e` instead of an `f`.

Like this:

Intended Address:

576f74506ff5e575fb1c4b42a145a8502977bee6

Mistyped Address:

576e74506ff5e575fb1c4b42a145a8502977bee6

One character off. Those bitcoin would be sent to a totally different destination.

Whose address is that? No one knows. It could be anyone's address in the entire world, or it could even be PKH that has no corresponding private key.

In the first case you'd make someone 5 bitcoin richer, and you have no way to reverse the transaction.

In the second case you would "digitally burn" those 5 bitcoin, sending them to a digital no man's land where no one can access them.

This would permanently reduce the total supply of bitcoin by 5 units and *ever so slightly* increase the value of all other bitcoin. Thanks for the donation to the network.

As you can see, typing out an address by hand is prone to errors and dangerous. Luckily, Satoshi invented something called Base58Check encoding to prevent this.

Hey, while you're inventing the world's first truly peer-to-peer electronic cash system that eliminates the need for third parties, you might as well also invent a way to prevent user typos in that system, too.

Thanks, Satoshi.

This typo-prevention process is step three of the address generation process, encoding the PKH with Base58Check.

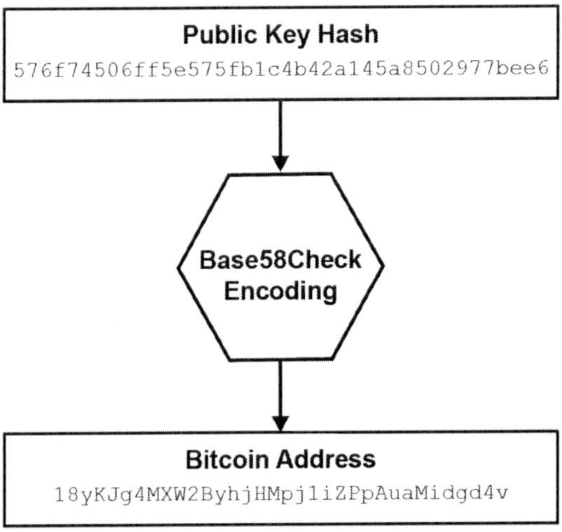


Here's how it works. Base encoding is a method used to represent large numbers in a more efficient way than hexadecimal. Hexadecimal requires two characters for each byte and sometimes isn't ideal.

Don't forget that our PKH is a very big number. In regular decimal format it's:

```
499,167,712,175,289,449,922,357,
955,396,465,327,006,367,071,974
```

Hexadecimal shortens this but base encoding does an even better job. Base encoding uses a mixture of alphanumeric characters to represent data. For example, base64 uses following character set:

```
ABCDEFGHIJKLMNOPQRSTUVWXYZabcdef
ghijklmnopqrstuvwxyz0123456789+/
```

Base58 is a subset of base64. It removes non-alphanumeric characters from the base64 characters set as well as ones that look similar and are often mistaken for each other. The base58 alphabet removes the plus sign, forward slash, lowercase L, capital I, capital O, and the number zero: `+/lIO0`. It also puts the numeric characters up front. The resulting base58 character set is:

```
123456789ABCDEFGHJKLMNPQRSTUV
WXYZabcdefghijkmnopqrstuvwxyz
```

With base64, encoding is very easy. `A` is encoded as `000000`, `B` is encoded as `000001`, `C` as `000010`, all the way up to the `/` symbol, encoded as `111111`. This makes a lot of sense. There are 26 capital letters, 26 lowercase letters, 10 numbers, and two special characters—64 different possibilities. Each character represents 6 bits.

With base58, there are only 58 different possibilities, so encoding is not as straightforward as base64. Each character represents less than 6 bits but more than 5 bits, so the process requires some additional steps.

Our starting point is our PKH:

```
576f74506ff5e575fb1c4b42a145a8502977bee6
```

First, we add what's known as the *version byte* to the PKH as a prefix. For a bitcoin P2PKH address it's `00`. The version byte helps identify the type of data we are dealing with. There are many different version bytes used in Bitcoin for different purposes. For example, the version byte for a P2SH address is `05`.

A PKH with a version byte added is known as a versioned PKH. In our example, it's:

```
00576f74506ff5e575fb1c4b42a145a8502977be
e6
```

The next step is to add a *checksum* to the versioned PKH. The checksum is 4 bytes of unique data that's added to the data from which it is generated. The checksum is used to prevent typing errors, which we'll discuss in a moment.

To obtain the checksum take the versioned PKH and hash it with double SHA256. The first four bytes of the resulting hash is the checksum. We take this checksum and append it to the end of our versioned PKH:

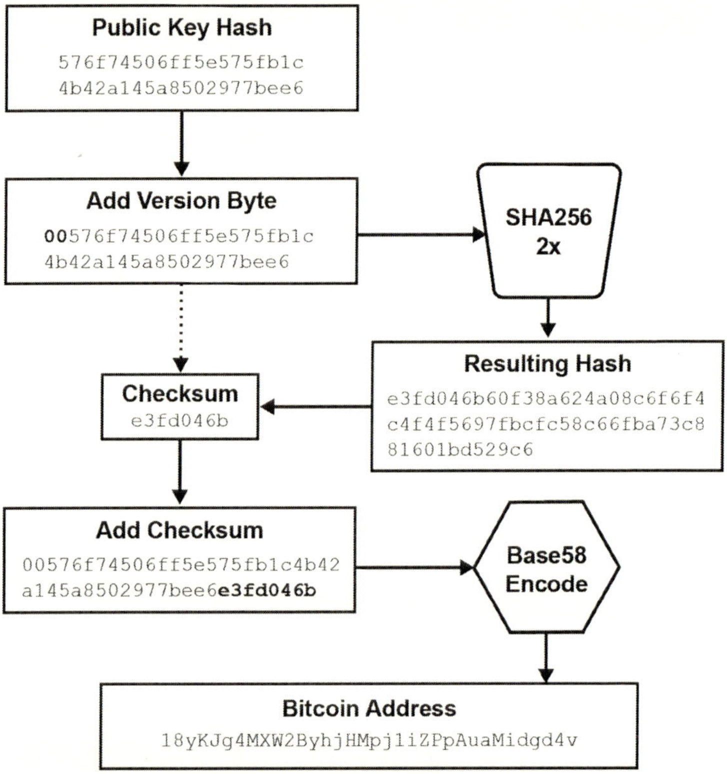


SHA256 Hash #1:

```
f0f2ae3e990f1a860b7a8b8fbd016593
fef643f8b8f91294c62b3e113e8c2357
```

SHA 256 Hash #2:

```
e3fd046b60f38a624a08c6f6f4c4f4f5
697fbcfc58c66fba73c881601bd529c6
```

Checksum: `e3fd046b`

Result: `00576f74506ff5e575fb1c4b42`
`a145a8502977bee6`**`e3fd046b`**

The result is a 25-byte hexadecimal number. This is what gets encoded in base58. Let's look at the actual encoding process.

The general idea of base58 encoding is to keep dividing our big number by 58 while keeping track of the remainders after each division. These remainders are what we use to encode the result using the base58 character set. But it's a little more nuanced than that.

The following character lookup table will come in very handy soon.

**Base58 Encode**
Character Lookup Table

| | | | | | | | |
|---|---|---|---|---|---|---|---|
| **0** | 1 | **15** | G | **30** | X | **45** | n |
| **1** | 2 | **16** | H | **31** | Y | **46** | o |
| **2** | 3 | **17** | J | **32** | Z | **47** | p |
| **3** | 4 | **18** | K | **33** | a | **48** | q |
| **4** | 5 | **19** | L | **34** | b | **49** | r |
| **5** | 6 | **20** | M | **35** | c | **50** | s |
| **6** | 7 | **21** | N | **36** | d | **51** | t |
| **7** | 8 | **22** | P | **37** | e | **52** | u |
| **8** | 9 | **23** | Q | **38** | f | **53** | v |
| **9** | A | **24** | R | **39** | g | **54** | w |
| **10** | B | **25** | S | **40** | h | **55** | x |
| **11** | C | **26** | T | **41** | i | **56** | y |
| **12** | D | **27** | U | **42** | j | **57** | z |
| **13** | E | **28** | V | **43** | k | | |
| **14** | F | **29** | W | **44** | m | | |

Here is the full process of encoding. Follow along, it's not difficult.

**Step 0**: Start with our (prefix + PKH + checksum) result.

```
00576f74506ff5e575fb1c4b4
2a145a8502977bee6e3fd046b
```

* * * * *

**Step 1**: Remove all leading `00` bytes and track them.

```
576f74506ff5e575fb1c4b42a
145a8502977bee6e3fd046b
```

Number of `00` bytes = 1

* * * * *

**Step 2**: Convert to a decimal number.

```
2,143,908,999,012,009,206,750,357,157,
633,245,293,490,292,157,903,433,172,075
```

* * * * *

**Step 3**: Divide by 58, track all remainders, and repeat until quotient is zero. The remainders are:

```
53 3 36 39 36 41 20 33 52 9 47 22 32 41
0 42 47 20 16 42 40 56 10 1 29 30 20 3
39 17 18 56 7
```

* * * * *

**Step 4**: Look up each remainder in the character lookup table and write down the corresponding character.

```
v4dgdiMauApPZi1jpMHjhyB2WXM4gJKy8
```

* * * * *

**Step 5**: Add the number of `00` bytes (in our case 1) to the end.

```
v4dgdiMauApPZi1jpMHjhyB2WXM4gJKy81
```

* * * * *

**Step 6**: Reverse the entire string of characters.

```
18yKJg4MXW2ByhjHMpj1iZPpAuaMidgd4v
```

And voila, the result of Step 6 is our bitcoin address. Now back to that checksum.

Before you send a transaction, Base58Check decodes the address, calculates the checksum and compares it to the checksum in the code. If they don't match, that means there's a typo in the address and the transaction is stopped. If they do match, the address is valid, and we can be sure that it's the intended destination of the bitcoin being spent in the transaction.

Here's how that happens. First the address is decoded by reversing the steps above. This gets us back to the starting point (prefix + PKH + checksum):

```
00576f74506ff5e575fb1c4b42a
145a8502977bee6e3fd046b
```

Next, the checksum, `e3fd046b`, is removed and noted. What remains is the versioned PKH:

```
00576f74506ff5e575fb1
c4b42a145a8502977bee6
```

This is then hashed twice using SHA256. The first 4 bytes of this double hash result *should equal the checksum.* If not, there was an error typing the address. Isn't that brilliant?

## ADDRESS COLLISIONS

As noted earlier, a bitcoin address is generated using RIPEMD160 which results in a 160-bit hash. This means that there are $2^{160}$ possibilities (approximately 1.46 x $10^{48}$). The possibility that two random individuals are assigned the same bitcoin address by chance is beyond remote.

Keep in mind that there is a separate private/public key pair for every bitcoin address. Your entire Bitcoin participation does *not* rely on a single private key. Your wallet will generate and store many key pairs resulting in many addresses for you.

If there ever happened to be an address collision, the other person could only access the bitcoin you have associated with *that* address, not your entire wallet. But again, the chance of an address collision is extremely, extremely low.

Every person on Earth could have the following number of bitcoin addresses before an address collision would be guaranteed:

190,000,000,000,000,000,000,000,000,000,000,000,000,000

You don't have to worry.

## DON'T REUSE ADDRESSES

Bitcoin is only pseudonymous. While difficult, potentially meaningful information and relationships between transactions can be uncovered. It's best practice to generate a new address for each incoming payment to maintain as much privacy as possible. Fortunately, most wallets will do this for you automatically.

* * *

# Break Time

*You've come far and learned a lot about the technical details of Bitcoin. Congrats! But there's still many important topics to cover. We'll continue the technical portion of the book with bitcoin wallets, blockchain, and Proof-of-Work. Once those topics are done, we'll finish Part 3 with some more details about transactions.*

# CHAPTER 10

# WALLETS

A bitcoin wallet is like the type of wallet that you keep in your pants pocket. But while a wallet in your pocket holds money, a bitcoin wallet does this and much more. It's a complex piece of software.

It's important to understand how you should *use* bitcoin wallets as a part of your overall Bitcoin investment strategy. This area of knowledge regarding bitcoin wallets includes the different categories of wallets, how to safely store wallets, backing up access to wallets, and more. We'll cover all of this in Part 4—Investing in Bitcoin.

In this chapter you'll learn what wallets are *doing* underneath the user interface. However, describing the full technical details and functionality of wallets is beyond the scope of this book.

Wallets are challenging for developers to do well and take a lot of know-how, so you don't need such specialized information to be discussed here. But, if it interests you, go deeper. It's a fascinating subject.

With that said, there are some technical items regarding wallets that I think you should know. Wallets are an integral part to Bitcoin investing, so you can't ignore them. You'll be much better off after learning a little bit about how bitcoin wallets work.

## WHAT WALLETS DO

At the highest level, a wallet allows a person to access their bitcoin via a user-friendly interface. Wallets look deceptively simple but the code running behind the scenes is complex. A wallet should at least have the following five major functionalities. Some wallets may have some more functionality, but these five capabilities are the baseline for a good wallet:

1. **Store Private Keys** - The primary thing a wallet does is store private keys. It's a common belief that wallets store actual bitcoin. This is not correct. All bitcoin units are stored on the Bitcoin blockchain. Wallets never contain any bitcoin. It's important to forever burn the following point into your brain: *private keys stored in the wallet only exist in the wallet and nowhere else. If you lose the wallet, you lose the keys and you lose your bitcoin forever.* Luckily, there are ways to back up your private keys, or more specifically your *seed phrase*. More on the seed phrase soon.

2. **Create and Manage Bitcoin Addresses and Keys** - A wallet should be able to create new Bitcoin addresses any time the user wants. This means that the wallet must generate a new a private key and public key first and store them while also keeping track of which key pairs correspond to which addresses.

3. **Display Your Addresses for Payments** - When someone wants to send you bitcoin, they need an easy and efficient way to obtain your address. Wallets need to display your preferred address quickly and cleanly. Copyable text is one way, but wallets often show

Quick Response (QR) codes that anyone with a smartphone can easily scan.

4. **Create and Sign Transactions** - The wallet should be able construct a transaction and digitally sign it using your unique Digital Signature. This is more complicated than it seems. You'll learn more about this in Chapter 13.

5. **Display Balances, Performance, and History** - Your wallet should show you how many units of bitcoin you have across all the addresses it tracks for you and their total current dollar value. It should display this information in real time, or as close to real time as possible. Your wallet might also show you the 24-hour percentage price change as well as month-to-date return, year-to-date return, and similar performance metrics. Finally, it should show you a history of your transaction activity within the wallet.

Bitcoin wallets are not distributed or shared on the Bitcoin blockchain. They are private and only you can see their contents. You control your wallet, and you are responsible for your wallet.

## NONDETERMINISTIC AND DETERMINISTIC WALLETS

The two main types of wallets are known as *nondeterministic wallets* and *deterministic wallets.*

Nondeterministic wallets generate each private key independently and randomly. There is no relationship between private keys. That's why these are also called random wallets. These were the first versions of bitcoin wallets available.

With nondeterministic wallets, each key must be backed up independently. This is a major disadvantage. When a new transaction occurs, a new backup should be done. This error-prone process is a hassle and therefore use of nondeterministic wallets is no longer recommended. Deterministic wallets have become the better solution for most Bitcoiners.

Deterministic wallets, also known as seeded wallets, are wallets where all the private keys are mathematically related to each other. In deterministic wallets all the keys are generated from one single random number known as the *random seed*. The entire wallet can be backed up simply by backing up the random seed. This also means that all the private keys saved in a deterministic wallet can be recovered with just the single random seed.

You only need to back up this random seed one time, when the wallet is created. This single backup is much more convenient than the frequent backups required with a nondeterministic wallet.

## HIERARCHICAL DETERMINISTIC WALLETS

Currently, the most advanced form of deterministic wallets is called a *Hierarchical Deterministic (HD) wallet*. HD wallets organize all keys into a tree structure. The seed generates a master private key that derives all other keys in the tree.

The master private key can generate child keys, which can be used to generate grandchild keys, and so on. There's no limit to the number of layers or keys that can exist in the tree. Also, each derivation path in the tree is unique.

First the random seed is hashed using the HMACSHA512 hashing function. This 512-bit hash is broken up into two halves. The left 256 bits become the master private key while the right 256 bits become what's known as the *chain code.* The chain code provides entropy when creating child keys.

Your master public key is created using your master private key as usual. The master private key plus the chain code equals your *master extended private key.* This is the first "parent key." From it, any number of child keys can be created. Each of those child keys can generate child keys themselves (effectively becoming their own parent keys).

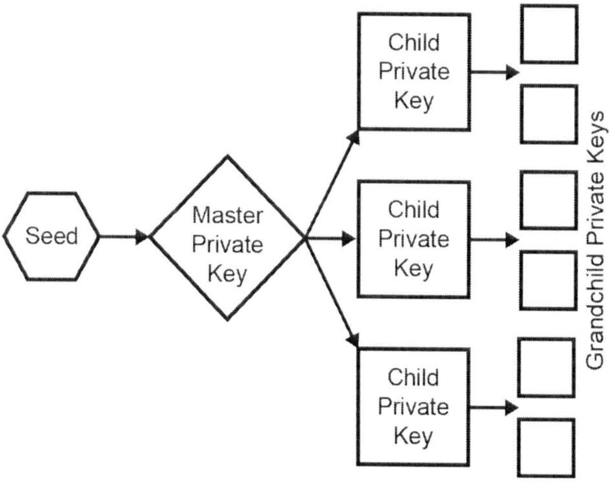


We won't go into the process of generating a child key from a parent key because it's a little complicated. The process includes using the following components: the parent extended private key, the parent private key, the parent public key, the parent chain code, an index, hashing, splitting hashes, adding hashes, and the child chain code.

The tree structure and way that the layers of keys are generated provide extra security and privacy while also allowing more organization. For example, say you derive three parent keys from your master extended private key, leading to three major branches of your tree. You could organize these branches into meaningful buckets such as "long-term investment", "child's savings", and "small bitcoin purchases" for example.

This is nice and could be useful. But the major advantage of an HD wallet is that the entire tree structure can be regenerated if you have the recovery seed phrase. You may wonder how this regeneration happens. The wallet doesn't generate an infinite number of addresses, that wouldn't work. Instead, wallets generate keys using standardized derivation paths. Wallets will generate address paths until they have reached a certain number of unused bitcoin addresses, *n*, in each layer. This number *n* is called the lookup gap or the gap limit. It's usually configurable in the wallet. *N* is not the same for all wallets. When it stops at *n* within each layer, you've got your initial tree structure back, plus some unused addresses ready to go.

This tree-from-seed capability makes it nice and easy to back up and store all your private keys in one shot. Because the seed phrase is so important, we're going to discuss it in detail.

## SEED PHRASE

The seed phrase, also called the recovery phrase, acts like a password but is much more secure. When you obtain a new wallet and open it for the first time, you'll see your seed phrase displayed. Often, it's a list of 24 short words. But depending on the exact wallet, it may be 18 or 12 words. The more words in your wallet's seed phrase, the more secure it is. A 24-word phrase is great, but even just 12 words is very secure.

I remember getting my first wallet. I saw my list of words and wondered what the heck it was. What did this list of words mean? How was it created? And how could a list of such simple words be so secure?

The answers are brilliant.

The set of words is a mnemonic for a big random number. Let's walk through an example of how a 12-word mnemonic seed phrase is created by a wallet. This same process applies to larger 18- and 24-word seed phrases.

There are six steps involved in making a mnemonic seed phrase:

1. Generate a random sequence of 128 bits (256 bits for 24 words).

2. Hash this sequence using SHA256. Take the first 4 bits (8 bits for 24 words) and use that as a checksum.

3. Add the checksum to the end of the sequence you generated in Step 1.

4. Split up your new sequence from Step 3 into 11-bit chunks.

5. Convert each 11-bit chunk to a decimal number.

6. Map each decimal to its corresponding BIP-39 word.

The result is your unique mnemonic seed phrase. The possibilities are so massive that no two people will ever have the same seed phrase.

I created a sample seed phrase just for this chapter. This type of word list is exactly what you'd see when opening a new HD bitcoin wallet for the first time. Of course, your words would be different:

```
electric blast assume carpet
curious depth segment since
traffic asthma disease planet
```

The following walks through how the seed phrase was generated step-by-step.

A random sequence of 128 bits is the same as flipping a coin 128 times and recording the result. In our example, my random sequence is (Step 1):

```
01000111011000101110110000110111
10010001011000110101111001110110
01110000110011100100100111100110
11000001110000000111111000101001 0
```

The checksum for this sequence is the first four bits of the SHA256 hash of this sequence. For this example, it's `1111` (Step 2). Thus, the sequence with the checksum added (Step 3) is:

```
01000111011000101110110000110111
10010001011000110101111001110110
01110000110011100100100111100110
11000001110000000111111000101001 0
1111
```

Next, break up the new sequence into 11-bit chunks (Step 4).

```
01000111011  00010111011  00001101111
00100010110  00110101111  00111011001
11000011001  11001001001  11100110110
00001110000  00111111000  10100101111
```

Each of these 11-bit chunks represent a decimal number between 0 (`00000000000`) and 2,047 (`11111111111`). In our case, these decimal numbers are (Step 5):

```
571   187   111   278   431   473
1561  1609  1846  112   504   1327
```

The last part of the process (Step 6) is to map each of these decimals to the corresponding BIP-39 word (BIP stands for Bitcoin Improvement Proposal). The list of BIP-39 words is 2,048 words long, numbered from 0 to 2,047. Each word is unique and doesn't repeat. Actually, the first four letters of each BIP-39 word are unique, making backup even easier (note that some words are only three letters long).

The entire BIP-39 word list can be found online but here is a sample of it.

| | |
|---|---|
| 0 | abandon |
| 1 | ability |
| 2 | able |
| ... | ... |
| 1000 | laptop |
| ... | ... |
| 2045 | zero |
| 2046 | zone |
| 2047 | zoo |

When mapping the decimals from Step 5 to the words in the BIP-39 word list, we arrive at our mnemonic seed phrase:

```
electric blast assume carpet
curious depth segment since
traffic asthma disease planet
```

Now when you start using a new wallet, you'll know exactly what the seed phrase means and represents. For a 12-word phrase, it represents 132 bits of entropy (264 bits for a 24-word phrase) from which your master extended private key is generated, enabling a whole tree of parent and child keys to be generated. See the Appendix for more on entropy and security strength.

* * *

That's enough discussion of wallets for now. Later in the book we'll talk more about wallets and how you can use them in your Bitcoin investment strategy. For example, should you use a software wallet or a hardware wallet? How do you back up and store your seed phrase?

We'll get there in Part 4 of the book. But next it's time to tackle a topic that's probably been on your mind for a while—blockchain. What exactly is the Bitcoin blockchain and how does it work? This revolutionary invention is next.

## CHAPTER 11

# BLOCKCHAIN

Blockchain. You've heard the word a million times but probably don't have a solid idea what it is. We'll change that. By the end of this chapter, you'll be a Bitcoin blockchain expert.

There's an enormous amount of mystique and confusion behind what blockchain is and how it's used. So, let's get one thing straight right off the bat. Satoshi invented blockchain for one purpose: to solve the issue of transferring digital assets between people without the need for an intermediary while also preventing double-spending. That's it folks!

Blockchain is extremely good at solving this problem because that's precisely what it was designed to solve. *Without blockchain, there'd be no Bitcoin. And if any third-party intermediary was required in Bitcoin at the network level, there'd be no need for blockchain.*

To understand Bitcoin, you need to understand blockchain. I won't leave you guessing. You're about to learn *exactly* what the Bitcoin blockchain is and how it works. I won't give you any vague, meaningless, or superficial descriptions here.

**Side Note:** *After this chapter and the next, you'll be armed with enough knowledge of blockchain technology to know its uses, potential, and limitations. You'll be able to call out people and* *businesses using the buzzword to try and impress you—which is often hilarious and always quite fun.*

## SIMPLE INTUITION

Bitcoin's blockchain is a distributed public ledger. Each full node in the network has a copy of it. This ledger contains the records of every single Bitcoin transaction since Bitcoin's launch on January 3rd, 2009. It continues to grow as new transactions are added to it.

Transactions, along with some other important data, are grouped into *blocks*. These blocks are then linked together forming a "chain of blocks" as Satoshi put it. He never actually used the term blockchain in his whitepaper. The term just stuck somewhere along the line.

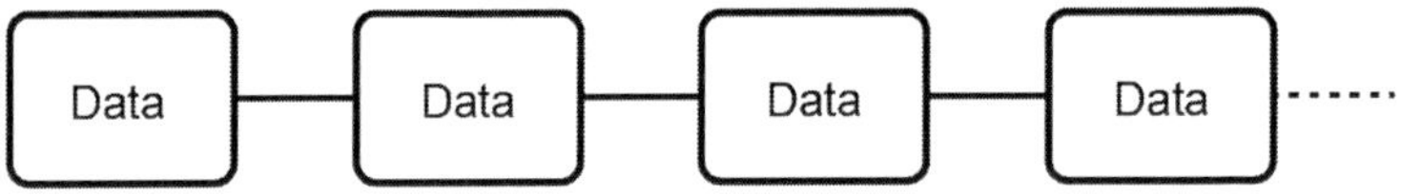


Blocks are a way of grouping transactions and permanently recording them so double-spending cannot occur.

Why use blocks?

Imagine that the network keeps track of every transaction that comes through in perfect order based on the split second each transaction was created. The network may receive a transaction at 12:05:00 pm, and another at 12:05:02 pm, and another at 12:05:03 pm, and so on.

The network would easily be able to tell whether a specific transaction was valid or not, that is, whether the bitcoin is available to be spent. It would simply have to compare every single new incoming transaction to all the transactions before it to see if the bitcoin is being double spent.

Sounds good, but Bitcoin does *not* do this. Since there is not one single central node controlling the network, this method can't possibly work. There are hundreds of thousands of transactions per day, many occurring at the exact same milli-second around the globe. With thousands of full nodes in the network, it would be impossible to reach exact consensus on the order and validity of all the transactions that come through in real time.

If instead of that method transactions were batched and processed once per day the validation process across all nodes in the network would be easier. This would prevent any ordering conflicts, but this method is not practical either. Every single transaction would be in total limbo for up to a full 24 hours without being validated and confirmed. That would be messy.

Bitcoin overcomes both problematic methods by using a process that occurs once every ten minutes on average. That's a very practical middle ground.

During that time a batch of transactions is verified, put into a block, and that block is added to the blockchain. This process also ensures that transactions are immutable, censorship is impossible, and the monetary value in the network is safe and secure.

To understand how all of this works and why these things are true, there are many aspects of Bitcoin's blockchain to cover.

After finishing this chapter and the next one you'll have a thorough understanding of:

- The contents of a block
- How a block is created
- How blocks are linked to each other
- How blocks are added to the blockchain
- How blocks are verified

Along the way we'll also cover some other relevant information not listed here. You're going to thoroughly understand this genius invention in no time.

## BLOCKCHAIN STRUCTURE

The very first block on the blockchain is called the *Genesis Block*. Each new block added to the chain is added in one direction, with the Genesis Block on one end, as time moves forward. *Each new block cryptographically references the previous block making the chain unbreakable.* On average, there are 144 new blocks added to the blockchain each day. This has been happening every day since Bitcoin launched and will continue every day forever.

Each block has a number. This number is also referred to as the *height* of the blockchain. The Genesis Block has a height of 0. As of the time of this writing, the Bitcoin blockchain has a height of 691,933 blocks.

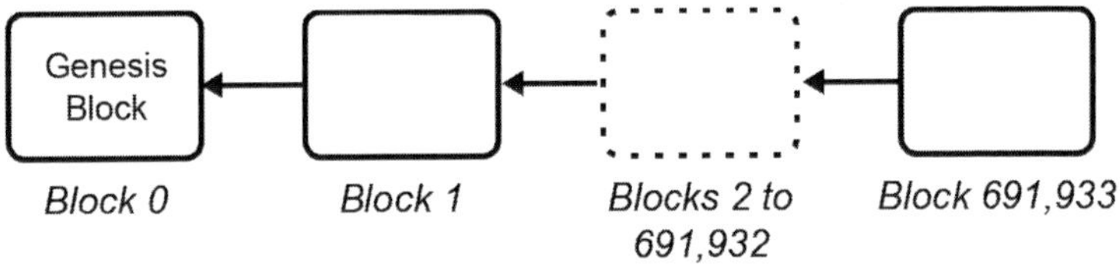


Satoshi added a secret message to the Genesis Block that references the headline of the British newspaper *The Times* on the day it was created, January 3rd, 2009. The secret message reads "The Times 03/Jan/2009 Chancellor on brink of second bailout for banks." It's speculated that this message serves to prove that the Genesis Block was created on this date but it's possible it was just included as a jab at the banking system.

Is it a clue that Satoshi hails from the United Kingdom or is it a red herring? We'll probably never know.

## STRUCTURE AND CONTENTS OF A BLOCK

Bitcoin blocks currently have a size limit of 1MB. There is some wiggle room for this due to something called *block weight*, but we won't be going into those details.

A block is just a data structure that can be thought of as a container for important information. The block contains four chunks of data:

- **Block Size:** A 4-byte field that tells us the size of the entire block in bytes
- **Block Header:** An 80-byte field that contains important block metadata
- **Transaction Counter:** A 1-to-9-byte field that tells us how many transactions are in the block
- **Transactions:** The actual transactions in the block (variable in total size)

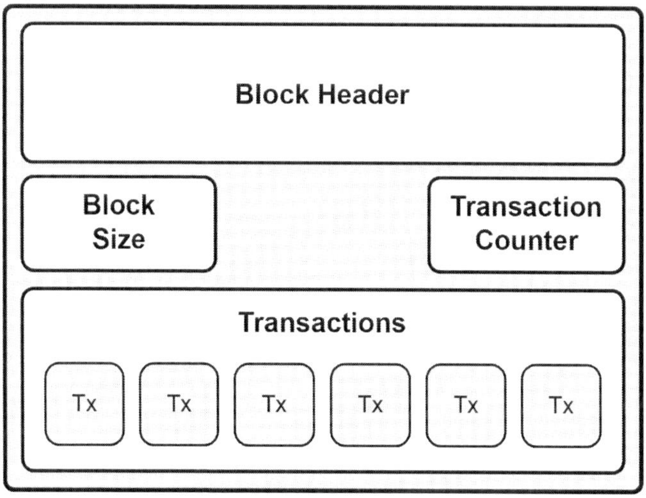

The Block Header contains six pieces of information, all a fixed size, that are critical to the blockchain's structure and operations. They are:

- **Version** (4 bytes): Block version information
- **Previous Block Hash** (32 bytes): The double SHA256 hash value of the previous block's header
- **Merkle Root** (32 bytes): A combined hash of all the transactions in this block
- **Timestamp** (4 bytes): The creation time of the block in Unix time
- **Target** (4 bytes): The calculated, network-determined Proof-of-Work target for this block
- **Nonce** (4 bytes): The counter used in the Proof-of-Work process that generates the solution for this block

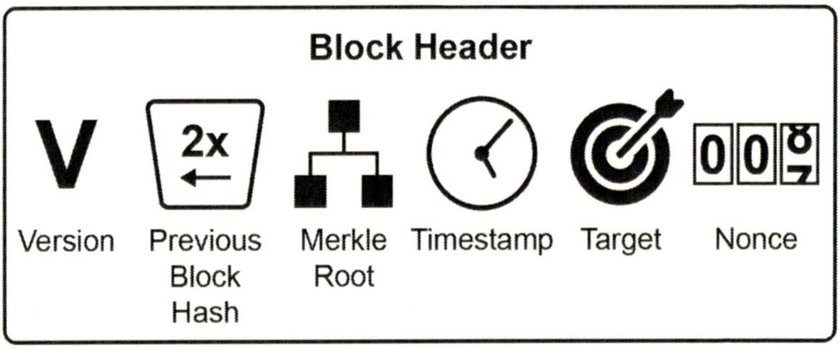


The first item, the version number, indicates which set of block validation rules to follow. There are currently four different versions in the wild. Version 1 was introduced in the Genesis Block and since then newer versions have emerged as Bitcoin has improved and evolved.

The timestamp marks the date and time that the block was created. It uses the Unix epoch (also known as Unix time or POSIX time) and describes the number of seconds that have passed since 00:00:00 January 1, 1970, UTC. This system of recording a point in time is widely used in computer operating systems. The timestamp also plays a role in Bitcoin mining.

The target and nonce are used in the Proof-of-Work process. We'll discuss these and the timestamp in more detail in the next chapter. For now, we'll focus on the Previous Block Hash and the Merkle Root.

## Previous Block Hash

The link from one block to the next in the chain is the double SHA256 hash of the block's header. To put it another way, the Block Header of the *current* block contains the double SHA256 hash of the *previous* block's header.

This Block Header hash of the current block is that block's unique identifier. This hash is not stored in the current block itself but is computed by nodes as they receive the block.

Thus, each block has a unique identifier that any node in the network can quickly calculate and confirm.

When a node receives a block from the network meant for its local copy of the blockchain, it needs to validate the block. As part of the validation process, it will compare the hash of the header of the *last* block on its blockchain with the Previous Block Hash included in the Block Header of the *incoming* block.

If these two hashes match and the rest of the validation succeeds the node will add the incoming block to its copy of the blockchain.

At that moment the blockchain gets one block and one link stronger.

## Merkle Root

A Merkle Tree is an efficient way to summarize large amounts of data while allowing the integrity of that data to be easily verified. Bitcoin uses Merkle Trees to summarize all the transactions that are in a block.

This process encodes all of a block's transactions (perhaps a couple thousand) into a single 32-byte hash. While Merkle Tree and Merkle Root may sound complicated and difficult, they're very easy topics for someone (like you) who already knows a lot about hash functions.

The process of finding the Merkle Root is simple.

First line up all the transactions in the block and hash each one using double SHA256. The transactions are ordered exactly as they are in the block Transactions section.

Next, start taking two transaction hashes at a time, concatenate (combine) them and hash the result using double SHA256.

For example, imagine you have a hash `aaaa` and another hash `bbbb`. Concatenating the two results in `aaaabbbb`. Then, `aaaabbbb` would be hashed.

Keep repeating this process of taking two child hashes at a time, concatenating them, and hashing the result to form parent hashes down the line. You'll now have a new layer consisting of parent hashes created by all of the transactions below. If there are *n* transaction hashes on the bottom layer, you'll have *n* divided by 2 parent hashes in the next layer.

Using this layer of parent hashes repeat the same exact process. Keep going up layers and eventually you'll wind up with only one single hash.

That's the Merkle Root.

Let's look at an example Merkle Tree using four transactions (abbreviated Tx) labeled A, B, C, and D.

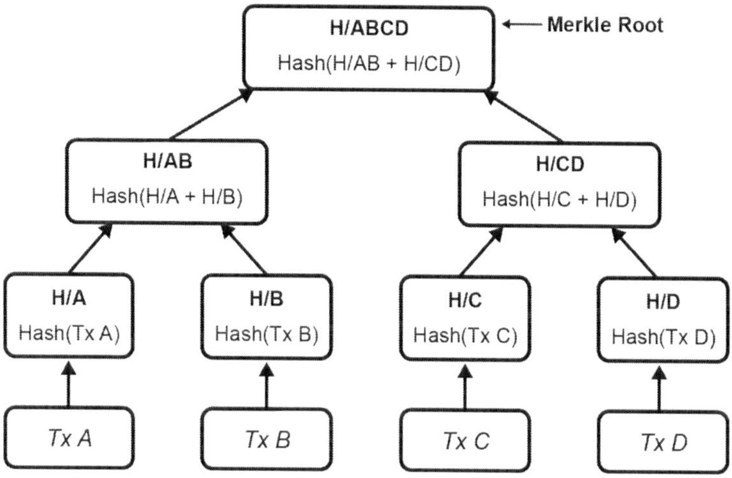


This process works the same way if there are hundreds or thousands of transactions in a block. No matter what, you wind up with a single 32-byte hash at the top of the tree. It works out nicely if there is an even number of transactions in the block. But what's done if there is an odd number of transactions? It's easy, the last transaction hash is duplicated.

Here's an example using transactions labeled A, B, and C.

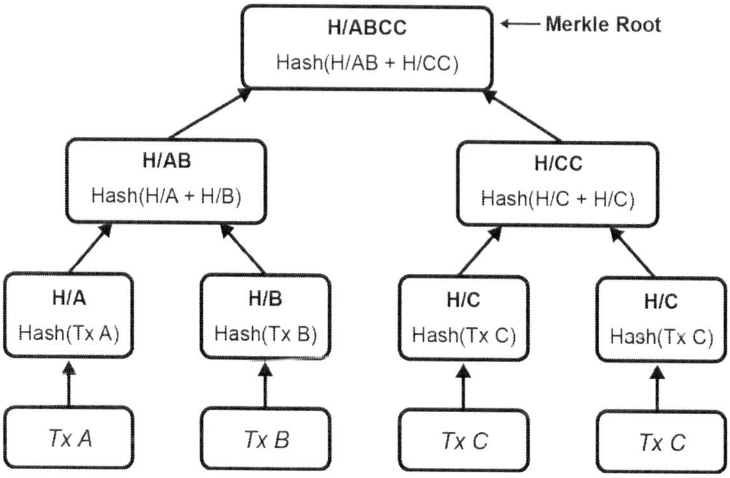

A Merkle Tree is simply a hierarchy of hashes that looks like an upside-down tree. The final hash in the tree is the Merkle Root that's included in the Block Header. The Merkle Tree in Bitcoin provides a way to show that transactions haven't been tampered with, proving their integrity efficiently and easily.

Placing a Merkle Root in the Block Header helps to make the blockchain immutable. If any transactions in a block are edited, added, or removed, the Merkle Root needs to be recalculated. This would change the Block Header and thus the block header hash. The entire digital fingerprint of the block would change, invalidating all subsequent blocks in the chain. The network would quickly notice, and nodes would not accept the change. We'll talk more about chain strength in the next chapter.

Merkle Trees also provide a way to verify transactions within a block without having to download the entire block. This is called Simplified Payment Verification (SPV), but we won't be going into those technical details.

## BLOCK CREATION

You've likely heard the term "Bitcoin mining" a trillion times. Though it's very straightforward, most people can't definitively tell you what miners do. It's common to hear that Bitcoin mining is comparable to gold mining. This makes it sound like Bitcoin miners' computers are digging through the dark corners of the internet uncovering shiny new bitcoin hidden in code. That's not the case at all.

Here's what Bitcoin miners do:

**Miners create new blocks to add to the blockchain and secure Bitcoin's decentralized network.**

Bitcoin miners do their work in five sequential steps. They are:

1. Validate individual transactions
2. Aggregate transactions
3. Construct the block
4. "Mine" the block
5. Transmit the block

### Validate individual transactions

Bitcoin wallets around the world are constantly creating new transactions and sending them to nearby nodes in the network. A node that receives a transaction independently validates it before passing it on to other nodes. Each of these nodes also validate the transaction, too.

In relatively little time, a single transaction has been received and validated by every node in the network. Invalid transactions are immediately dropped from the network by the first node that encounters them.

The transaction validation process that a single new transaction goes through at each node along the way includes a list of criteria that the transaction must pass. Some criteria are obvious, such as ensuring the syntax and data structure of the transaction are correct. Other criteria are a bit technical, such as "the unlocking scripts for each input must validate against the corresponding output locking scripts."

The main purpose of transaction validation is to make sure that the transaction follows the rules of the Bitcoin network.

When a node validates a single transaction, it sends it to other nodes but also puts it into a group of valid but unconfirmed transactions called the Memory Pool, or *mempool*.

**Aggregate transactions**

In this step, miners look through the mempool and choose which transactions to include in their *candidate block.* A candidate block is a block that's put together but not yet validated.

Miners want their blocks to be smaller in size for increased speed in the mining competition, so they must be selective when choosing transactions.

The mining process is incentivized by the *block reward,* the financial reward miners earn for successfully creating a block with a valid Proof-of-Work. It consists to two components, the *block subsidy* and transaction fees.

The block subsidy is the quantity of newly minted bitcoin that the miner earns for the block. When Bitcoin was initially released miners earned 50 newly minted bitcoin for each block mined. But it's programmed into Bitcoin that every 210,000 blocks the subsidy gets cut in half. This halving will continue until the very last (the 21,000,000$^{th}$) bitcoin is mined. The most recent halving was on May 11$^{th}$, 2020. Currently, the block subsidy is 6.25 BTC.

Transaction fees are also a component of the block reward and transactions with higher fees are naturally more attractive to miners. Fees fluctuate in the network but are low. However, almost certainly, the higher the fee you decide to pay the faster your transaction will get put into a block. After the final bitcoin is mined the network will still run and miners will still perform their work, but it will be for transaction fees only.

In summary, during this step miners select a set of transactions that give them decent transaction fees while also keeping the total block size reasonable, so they have a good chance of winning the block reward. If a miner gets *too* picky about constructing their candidate block, they may lose precious time and be beaten by another miner.

## Construct the block

Once the miner has selected the transactions to include in the candidate block, he will construct the first transaction in that new block, its coinbase transaction. This transaction includes the block subsidy and transaction fees. The coinbase transaction sends this money to the miner's bitcoin address. All other transactions in the block follow the coinbase transaction.

The miner then constructs the Block Header. This includes the six components discussed earlier. The two components we haven't discussed in detail yet are the target and the nonce.

The Bitcoin network has a fluctuating difficulty target for its Proof-of-Work. The miner grabs the current target from the network. All miners are always using the same target as everyone else. The nonce is just a number used as a counter. It's initially set to zero.

Once the entire Block Header is prepared the block is ready to be mined.

## Mine the block

Once the block is prepared the miner begins using his computational power to mine it. *Bitcoin mining is finding a valid Proof-of-Work solution to the candidate block.*

The details of Proof-of-Work are covered in the next chapter, but the general idea of mining is easy to explain. The Block Header is hashed. That hash is checked to see if it meets the target. If so, the miner has a valid block. If not, they try again with a new hash.

Once a block is validated with Proof-of-Work, it's ready to be transmitted to the rest of the network. Transactions technically aren't confirmed until they are included in a mined block.

### Transmit the block

The miner lets the world know that he has a valid block by transmitting it to other nodes in the network. These nodes validate the block themselves, accept it, and spread it across the entire Bitcoin network.

Recall this mining process takes roughly 10 minutes per block. Imagine that the blockchain is at a height of 900,000 blocks. There could be thousands of miners all working simultaneously on trying to find a solution to block 900,001. The miners are all hashing away when suddenly, a valid block 900,001 propagates across the network. Immediately, all the miners stop trying to find a valid block for height 900,001 and get right to work on trying to find a Proof-of-Work solution to block 900,002.

No time is wasted.

* * *

There's more to learn about Bitcoin's blockchain but first you need to learn more about the target and nonce. From there, we can discuss the details of Bitcoin mining and cover some other interesting blockchain concepts.

So, let's not twiddle our thumbs. Onward to the next chapter! Proof-of-Work is next.

## CHAPTER 12

# PROOF-OF-WORK

*"Proof-of-work has the nice property that it can be relayed through untrusted middlemen. We don't have to worry about a chain of custody of communication. It doesn't matter who tells you a longest chain, the Proof-of-Work speaks for itself."*

- Satoshi Nakamoto, August 7th, 2010

Bitcoin is all about proof and verification. It's actually *because* of Bitcoin's massive reliance on proof and verification that the need for trust between parties is removed from the equation entirely. This chapter explores more about this and details Proof-of-Work, which is critical to Bitcoin.

As you know, all transactions are validated by each node as they propagate through the network. They're checked for syntactical accuracy, whether the sender has the funds, and bunch of other criteria. If a transaction doesn't meet the strict criteria, it's not executed and is dropped from the network, never to be seen again. Valid transactions continue and get grouped into blocks to be added to the blockchain.

Each full node also has a copy of Bitcoin's blockchain that it updates as new blocks arrive. It's crucial that nodes agree on the current valid state of the blockchain, so everyone is on the same page regarding the status of individual transactions across the entire network.

As a new block is created and presented to the network, nodes go through a process of verifying it just like they do for individual transactions. Once a majority of nodes in the network verify a block and, in a sense, "vote" for it, the block is added to the blockchain, and the miner of that block receives the block reward.

The moment a block is added to the blockchain and linked to the previous block, it's as if the network says: "these transactions are completed, accounted for, and recorded. Let's continue with the next batch of transactions and next block."

When a miner presents a block to the network to be added to the blockchain the block's Proof-of-Work solution must be included. This Proof-of-Work is the final stamp on the block and basically means that "these transactions are ready for the blockchain and here's the proof."

Finding this valid Proof-of-Work can only be accomplished by expending energy and using computational power. There's no analytical way to find a solution so there's no possibility of any miner having an intellectual edge over any others. Finding a solution must be done by brute force.

Therefore, it's called Proof-of-Work. There's absolutely no way to fake the effort required to find a solution to the block. Miners really do have to work for it. When a valid Proof-of-Work is presented everyone in the network can be certain that it was found by converting energy and processing power into work.

An important idea underlying Proof-of-Work is that it's difficult to find a solution but easy to verify that solution once found. It takes a lot of energy, computation, and work to build the blockchain, but effectively zero energy, computation, and work to verify the information in the blockchain. This is the magic that gives the blockchain integrity, security, immutability, and censorship resistance.

Say an attacker wants to edit a transaction from 6 blocks back in the chain to financially benefit himself. If he tried, he'd break all links in the chain from that block forward. Remember that a block is cryptographically linked to the previous block through the Block Header hash and changing any single piece of information in a block changes the Block Header hash. Any attempted edit automatically causes the links to break.

To succeed, the attacker would need to make their malicious edit, then find Proof-of-Work solutions for that block *and the next five blocks* all while keeping up with the current additions to the chain. By the time they would find solutions to these 6 blocks, the rest of the network is likely focused on a version of the blockchain that is 5 or 6 *more* blocks ahead than when the attacker started.

Not only does an attacker have to find new Proof-of-Work solutions to those blocks, but he also must do it faster than the rest of the network and then get a majority of other nodes in the network to "vote yes" on his new blocks and version of the blockchain. This just isn't going to happen. No single miner in the world controls that much mining power. Even if a single miner combined all the supercomputers in the world in a concentrated effort, he still couldn't tip the odds in his favor—the Bitcoin network's total hashing power is much greater.

The blockchain's structure and design leads to an interesting feature. The enormous cost in terms of electricity and computing power combined with the near-zero probability that an attack can be done fast enough means that an attacker is *way better off focusing on new legit, valid blocks. Trying to attack the network is a guaranteed monetary loss. Playing fair has positive expected value.*

All nodes can independently verify that the network is "correct and OK" at any time, quickly and for free. But the cost of adding new blocks to the blockchain is expensive and continues to rise.

Part of Satoshi's genius was designing a system where a foundational assumption is that all nodes are potential attackers who can't be trusted. Since no one can or should be trusted he removed trust as a factor from the network entirely. There's no need for any participants in the network to build trust between each other.

The highly asymmetric costs of producing a valid Proof-of-Work versus verifying it keeps all nodes honest, playing fair, and gives them skin-in-the-game regarding the success of Bitcoin as a whole. The incentive to defraud the system with a fake transaction is eliminated since it's far too expensive and has a ridiculously low probability of success.

## EMERGENT CONSENSUS

Consensus is how participants in a system who have shared goals reach an agreement. In Bitcoin, consensus is reached across the decentralized network on the correct state of the blockchain through transaction verification, valid Proof-of-Work for individual blocks, verification of these blocks and their linking into a chain, and finally by nodes selecting the state of the blockchain with the most demonstrated amount of Proof-of-Work—the chain that has the most demonstrated and cumulative computational power and energy put into it.

There is no vote per se, and no precise point in time when all the nodes agree on a particular block. But consensus rather emerges as thousands of nodes in Bitcoin's network all continue to operate while following the same rules.

If a block is valid, a node accepts it. But let's not forget about all the cryptographic beauty that gets us to the point where a block is created and ready for validation. Transactions from around the world are created by bitcoin wallets using hash functions, digital keys, digital signatures, and bitcoin addresses. And each block of transactions contains a version number, the previous block's hash, a Merkle Root of all the transactions, a timestamp, a target, and a nonce.

When more than 50% of nodes accept the block and include it in their local copies of the blockchain, consensus is reached. Nodes will always follow the *longest chain rule* which states that nodes must follow the chain with the most cumulative work, the chain that took the most effort to build.

This emergent consensus is what prevents double-spending. Proof-of-Work plays a critical role in this Bitcoin phenomenon, so let's look at exactly what it is.

## THE PROOF-OF-WORK TARGET

As mentioned in the previous chapter, Bitcoin mining is the process of finding a valid Proof-of-Work solution to a candidate block. Mining keeps the network safe and secure, keeps it censorship resistant, and creates durable, immutable, truthful records of ownership. Nodes that successfully mine a candidate block are rewarded with newly minted bitcoin and transaction fees paid by network users.

Mining takes electricity and computational power. It's a mechanism for efficiently converting raw energy into monetary value and network security. In this sense, it's one of the most efficient technologies that humans ever created.

You may have heard the description that Bitcoin mining is "solving highly complex mathematical puzzles." This makes mining sound almost Terminator-ish.

That description of mining is incorrect and misleading. Mining is not mathematical wizardry. It's just hashing. More specifically, it's finding a suitable hash for a candidate block. For a block to be accepted by the network, its hash must meet a criterion. Even though a miner has all the data from within a block, she can't just hash it and move on. *The output hash might not be good enough.*

Why?

Satoshi shows more of his genius here. Recall that in the Block Header of each block there's a target and a nonce. We have not discussed these two fields of the header yet. Let's talk about the target now and come back to the nonce in a moment.

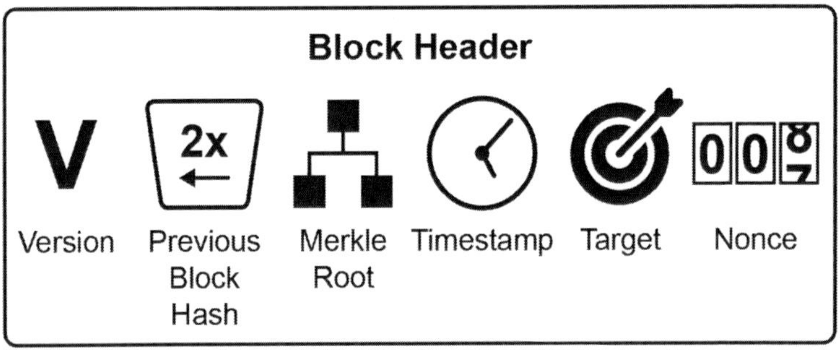


The target is a predetermined 256-bit number that's agreed upon by the entire Bitcoin network. **A valid Proof-of-Work solution is found when the Block ID (the hash of the Block Header) is less than or equal to the target.**

The target changes every 2,106 blocks. That's roughly every two weeks. The target fluctuates up or down to make bitcoin mining easier or more difficult as needed. The approximately 10 minutes between blocks being added to the blockchain is consistent because the target is adjusted regularly. Satoshi himself said the target is there "to compensate for increasing hardware speed and varying interest in running nodes over time."

If blocks are generated too fast the difficulty of mining increases, slowing down block generation. If blocks are generated too slowly the difficulty decreases, speeding up block generation. This periodic adjustment creates a smooth and totally predictable schedule for the issuance of new bitcoin to the Bitcoin money supply.

The maximum possible target for Bitcoin Proof-of-Work is:

```
26,959,535,291,011,309,493,156,
476,344,723,991,336,010,898,738,
574,164,086,137,773,096,960
```

This decimal number converted to hexadecimal is:

```
00000000ffff0000000000000000000000
0000000000000000000000000000000000
```

Recall that the SHA256 hashing algorithm has $2^{256}$ possible outputs. Just to remind you, that number in decimal form is:

```
115,792,089,237,316,195,423,570,985,
008,687,907,853,269,984,665,640,564,
039,457,584,007,913,129,639,936
```

Imagine that you took a small piece of paper and wrote 0 on it, crumbled it up into a ball about the size of a pea and set it aside. You do this again with the number 1. Then again with the number 2. And again, all the way up to $2^{256}$. Next you put all those crumbled up pieces of paper into a hat.

(That hat would need to be as large as approximately 136,740,776,142,319,550,571,057 Milky Way galaxies.)

Now shake the hat to mix up the numbers. Once they are thoroughly mixed stick your hand into the hat and pick out one crumbled up piece of paper. Chances are that you'd select an enormous number. It would probably be somewhere around the average of the numbers in the hat, approximately $5.789 \times 10^{76}$. The *maximum* target is *really* small compared to $2^{256}$.

Since miners are looking for a hash value that's lower than or equal to the target, keep in mind the following:

- $0 \le$ Solution $\le$ Target
- The lower the target, the more difficult the task of finding a solution
- The higher the target, the easier the task of finding a solution.

As of the time of this writing, the current actual target used in the Bitcoin network is:

```
0000000000000000000d5f7b00000000
00000000000000000000000000000000
```

This number is *way* smaller than the maximum target. In decimal form, the current target is:

```
1,280,876,111,474,813,957,445,809
,627,925,710,317,539,675,495,922,
139,136
```

We'll come back to this current target shortly. For now, let's finish up discussing the mining process.

## BITCOIN MINING SUMMARIZED

When constructing the Block Header for their candidate block, a miner already has the required target since it's predetermined by the Bitcoin network for a given time period. The final piece of the puzzle is the nonce. It's the last part of the Block Header that you need to learn, and it's not complicated at all.

A nonce is simply a counter. There isn't anything special to it. However, it *is* a piece of data. Changing the nonce changes the Block Header hash. All else equal, a Block Header with a nonce of 0 will have a totally different hash than a Block Header with a nonce of 1, or 2, or 3, etc.

So, armed with your new knowledge of the nonce, here is the summarized process that a Bitcoin miner goes through when mining:

1. Validate and aggregate individual transactions.
2. Construct the Block Header and use the network's current target.
3. Set the nonce to 0.
4. Hash the Block Header using double SHA256.
5. Check the hash. If it's less than or equal to the target, congratulations! You've got a valid Proof-of-Work solution.
6. If not, increment the nonce by 1, rehash, and compare again. Keep incrementing the nonce until you find a hash that's less than or equal to the target.

Satoshi's system is genius. Miners need to find the correct nonce that makes the hash acceptable. This repeated loop of "try a nonce, check the resulting hash" until finding a hash that locks the block can *only* be done by brute force trial and error. There's no analytic solution to it. Because of the properties of cryptographic hash functions described in Chapter 6, miners must keep guessing until they find a solution.

Finding a solution takes pure energy and computational power; there's no shortcut. When a miner shows the network that they have a hash for the block that's below the current target, it really is proof that they converted energy and computation power into a solution. There's no other way to find it.

*Side Note*

You might be wondering, are all the miners in the world simultaneously working on the same exact puzzle?

No. Each miner is working on a unique puzzle.

It's unlikely that two miners would select the same exact transactions, in the same order for their block (which changes the Merkle Tree). And if this did happen, and they timestamped their blocks at the exact same second, it still wouldn't matter because of the coinbase transaction.

Remember that the coinbase transaction sends the block reward to the miner's bitcoin address. No matter if everything in the block is the same as other miners, the Block Header would still be different because the miner's address affects the hash.

Individual miners are always solving unique puzzles.

## EXAMPLE VALID PROOF-OF-WORK

Miners are just ticking through nonces until they find one that works. For example, pretend that you created a candidate block and set the nonce the 0. Then you hashed it, resulting in your hashed block ID. Let's assume the following is what popped out:

Nonce: 0

Block ID:
```
0000000031ff925e4cf8
24b8d714096f600d0595c
9df19a0c988bde04d63000
```

Target:
```
000000000000000000d5
f7b000000000000000000
0000000000000000000000
```

Your block's hash is much larger than the target. It's about 257 billion times bigger. You don't have a valid Proof-of-Work yet. So, you increment the nonce to 1, and try again. The block's new hash versus the target is:

Nonce: 1

Block ID:
```
0004206279744aae44205
80ac5132c9d5e7ab0f947
232eedc7e1262152ff3013
```

Target:
```
000000000000000000d5
f7b000000000000000000
0000000000000000000000
```

Ouch. That got much worse. Now the block hash is over 5.6 quintillion times as large as the target. So, you keep trying hashes, incrementing the nonce by 1 each time. You just never know what nonce is going to work. Say you're up to a point where the nonce is now 212,137,086. And this is what you get:

Nonce: 212,137,086

```
Block ID:  0000000000000000000
           8627fafbf04947356a9fa
           2a394ba381545a0c7762fb

  Target:  00000000000000000000d5
           f7b000000000000000000
           0000000000000000000000
```

Bingo! You've found a block ID that is lower than the target. It took you over 212 million tries, but you finally have a valid Proof-of-Work solution. Now you can let the whole Bitcoin network know that you've got a block ready for the blockchain and are able to prove it. Immediately, other miners will verify the block. Once enough of them do, the block is get added to the blockchain and all the miners begin working on the next block.

## TARGET BITS

The `nBits` field in a block's Block Header is a compact way to store the target. Since we're dealing with a constantly growing global network, every bit and byte in a block matter. Real estate is competitive. Instead of storing target as a 64-digit hexadecimal number, it's stored in the header as 4 bytes.

Recall that, as of the time of this writing, the current target used by all miners in the network is:

```
0000000000000000000d5f7b00000000
00000000000000000000000000000000
```

The bits of this target are `0x170d5f7b`. (Remember, that `0x` in the beginning just signifies that the following number is in hexadecimal.)

There's a formula to apply to the bits, whatever they are for a given period, to arrive at the big 64-digit hexadecimal target. It is:

$$\text{target} = \text{coefficient} \times 2^{(8 * (\text{exponent} - 3))}$$

The first byte from the bits field, `17` in our example, is the exponent. The last three bytes, `0d5f7b`, are the coefficient. When we plug these values into the formula we get:

$$\text{target} = \texttt{0x0d5f7b} \times 2^{(8 * (\texttt{0x17} - 3))}$$

$$\text{target} = \texttt{0x0d5f7b} \times 2^{(8 * (23 - 3))}$$

$$\text{target} = \texttt{0x0d5f7b} \times 2^{(160)}$$

$$\text{target} = 876{,}411 \times 2^{(160)}$$

That calculation results in the massive number:

```
1,280,876,111,474,813,957,445,809,627
,925,710,317,539,675,495,922,139,136
```

This large number in hexadecimal is exactly the current target used:

```
0000000000000000000d5f7b00000000
00000000000000000000000000000000
```

## DIFFICULTY

It's very easy for computers to read large hexadecimal numbers but difficult and cumbersome for humans. Lucky for us, a way to represent the target in a more human-friendly way was devised. It's called *difficulty*. By quickly looking at difficulty values a person can compare any different hexadecimal targets much easier.

$$\text{difficulty} = \texttt{0xffff} \times 2^{(8 \times (\texttt{0x1d} - 3))} / \text{target}$$

The bits of the maximum possible target are `0x1d00ffff` and they can be seen in the above equation. The difficulty of the *maximum* possible target is 1. Using the current actual target in Bitcoin and this formula, we arrive at a current difficulty of 21,047,730,572,451. While that's a big number, it's much easier for us mere moral humans to read than a target's hexadecimal.

The difficulty is interpreted as *how difficult it is to mine the current block compared to the genesis block*. Currently, mining a block is 21,047,730,572,451 times more difficult that it was to mine the very first block.

## RETARGETING / DIFFICULTY ADJUSTMENTS

The target exists to regulate the time interval between successfully mined blocks and create a consistent amount of time between blocks being added to the blockchain. This results in predictable growth in the money supply and predictable transaction settlement times.

Every 2,016 blocks the target is adjusted. The actual adjustment is called a retarget or difficulty adjustment. The retarget is done automatically by each node in the network—it's built into the rules of Bitcoin that all nodes follow.

Here is the formula used:

$$\text{Target} = \text{Old Target} * (\text{Actual Time}_{\text{last 2,016 blocks}} / 20{,}160)$$

There are 20,160 minutes in a two-week period. This is the expected amount of time to mine 2,016 blocks. The actual time it takes to mine the previous 2,016 blocks is divided by this expected amount of time, resulting in a target change factor. This target change factor is then multiplied by the current target to arrive at the new target for the next 2,016 blocks.

The above formula is a summarized version. The target change factor can be no higher than 4, and no less than 0.25 per adjustment period. These limits are to avoid too much volatility in targets from one period to the next.

But the general idea of retargeting holds true. If the last 2,016 blocks took longer than two weeks to mine, then the target needs to increase (meaning difficulty must decrease). If it took less than two weeks to mine the last 2,016 blocks, then the target needs to go down (difficulty must increase).

## MORE ON THE NONCE

A nonce is slightly more than just a counter. It's a cryptographic "number used only once", hence the name nonce. Nonces aren't unique to Bitcoin. They're used a lot in different software programs.

In Bitcoin, the nonce is only a 32-bit (4 byte) number. This means that there are only $2^{32}$ different nonces to choose from. That's 4,294,967,296 nonces.

That's not a big number when compared to all other numbers in Bitcoin. Because of its small size limit, there may be times when a miner tries all possible nonces and still can't find a valid Proof-of-Work solution. There are five different things miners can do in this situation.

The easiest option is to change the timestamp in the Block Header. The other four options involve transactions. The miner could: modify the coinbase transaction, add one or more regular transactions, remove one or more regular transactions, or simply rearrange the current transactions. Making edits to the block using any of these five options changes the Merkle Root.

Regardless which adjustment the miner chooses to perform the Block Header will be new. It's time for him to try those 4,294,967,296 nonces all over again. This process can be repeated again and again. A 32-bit nonce is plenty.

## BLOCK VALIDATION AND TIMESTAMP RULES

Validation also occurs with blocks. Nodes validate them before passing them on to other nodes, just like individual transactions. Valid blocks spread through network quickly, invalid ones are rejected.

Miners whose blocks are rejected lose their block reward and just waste a lot of money on electricity. Again, there's no incentive to cheat.

A node that receives a new block checks it against a variety of criteria. Some of the more important ones include ensuring the Block ID is less than the target, the timestamp is within the acceptable range, the coinbase transaction is structured correctly, and all other transactions in the block are valid. If these and all other criteria are met the block is considered valid and passed along to peers.

It's worth mentioning the rules for timestamp validation here. The blocks in the blockchain *don't* have to be timestamped in perfect sequential order. Consider the following example:

- Block 900,000 - 07:15:37 am
- Block 900,001 - 07:25:55 am
- Block 900,002 - 07:34:05 am

- Block 900,003 - 07:28:11 am
- Block 900,004 - 07:46:09 am

Even though Block 900,003 has a timestamp that's earlier than Block 900,002 that's OK if the timestamp meets two specific criteria. First, the timestamp of a block must be later than the median timestamp of the previous 11 blocks. Second, the timestamp must be less than the network-adjusted time plus 2 hours. The network-adjusted time is the median timestamp of all the nodes connected to you.

*These timestamp rules exist so no one can manipulate or influence the next target calculation.* Specifically, this prevents miners from manipulating the target so the difficulty intentionally *decreases* next adjustment. If a much older block was accepted onto the blockchain the target change factor could easily be maxed out, resulting in the target going up by a factor of four, making mining much easier for the next two weeks.

This same type of manipulation could happen again and again. Soon enough, the target would reach its max. Considering the formidable mining power currently in the network blocks would be mined much quicker than every 10 minutes and concentrated mining power could have an unfair advantage in earning new bitcoin.

## BLOCKCHAIN FORKS

It's possible for different copies of the blockchain to exist at the same time. This is called a *blockchain fork* or blockchain split. It happens when two miners produce a new block for the blockchain at roughly the same time.

Try to imagine the huge global decentralized Bitcoin network like a web around the world. And for simplicity's sake, let's assume that every single node in the network agrees that the blockchain is currently on block 900,000.

Let's say that a miner in Cleveland, Ohio, finds a Proof-of-Work solution to block 900,001. She broadcasts it and nearby nodes begin validating that block, spreading it to other nodes in the network. The Cleveland version of the blockchain continues to spread.

At roughly the same time, perhaps just a split second after the Cleveland miner found a solution, a miner in Tokyo, Japan, also finds a solution to block 900,001. While the Tokyo miner was mining his block 900,001, the Cleveland block 900,001 had not reached him yet. So, the Tokyo miner broadcasts his block to nearby nodes.

Now we have a situation where the Cleveland version of the blockchain and the Tokyo version of the blockchain, both valid, are spreading across the entire network. Perhaps half of the nodes accept the Cleveland version and half of the nodes accept the Tokyo version.

Now we have two blocks at height 900,001. Both blocks are valid, both have a Proof-of-Work solution, and both blocks cryptographically link back to block 900,000. It's very likely that the two blocks have some of the same transactions in them, too. We just don't know which block is the winner yet. But we will soon.

**Competing Chains**

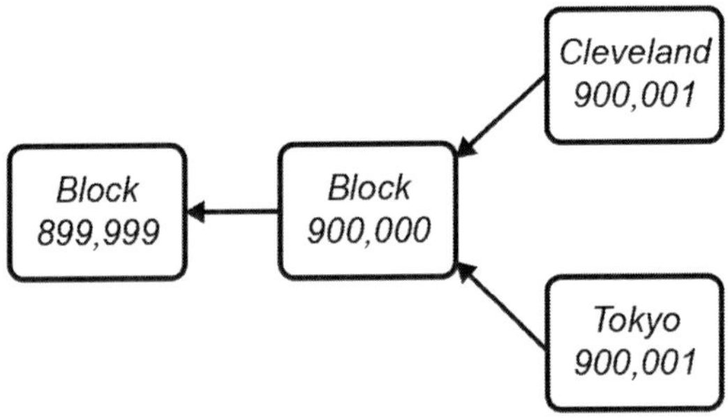

At this point there are two equally valid versions of the Bitcoin blockchain. A fork has occurred. By now, both the Cleveland block and the Tokyo block have spread completely around the network and every node in the network has a copy of one of the versions of the blockchain.

Nodes that have accepted the Cleveland version of the blockchain will also receive the Tokyo block within a short period of time. They consider it a losing block, but don't delete it. They still link it to block 900,000 in their copy of the blockchain, just like they did the Cleveland block, forming another chain. They hang on to the Tokyo block for a little while in case it turns from a loser to a winner. This same thing happens for nodes that initially accepted the Tokyo version on the blockchain and receive the Cleveland block.

Now everyone in the network is waiting for the next mined block. About 10 minutes go by and a miner in London, England finds a solution for block 900,002. Here's the question that matters: what version of the blockchain is the London miner on?

Let's assume he is on the Tokyo version of the blockchain. Thus, the miner is linking his mined block 900,002 to the Tokyo version block 900,001.

London broadcasts his new valid block. This mining period, there are no conflicts at all. The entire Bitcoin network sees London's block 900,002 and accepts it. It's at this moment that we know that the Tokyo version of the blockchain wins. Cleveland loses. All miners in the network switch over to the Tokyo-London version of the blockchain and discard the Cleveland remnants.

The Cleveland block becomes what's known as an *orphan block*.

All valid transactions from the Cleveland block that were not included in the Tokyo block go right back into the mempool and will most likely be included in the next block.

**The Tokyo Block Wins**

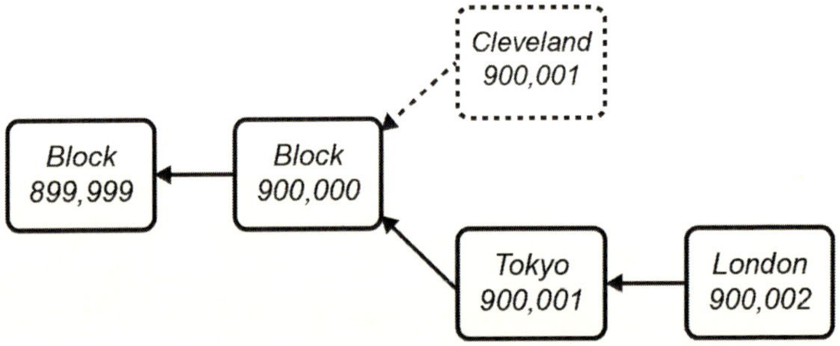


Miners will always converge (or even reconverge) onto the blockchain version that has the most demonstrated work. The chain with the most cumulative work always wins.

For the most part, forks are resolved within one block. We won't go into the details of how to calculate the probability of a fork occurring but for those interested, block finding is an "inhomogeneous Poisson process."

But it's useful to know that the probability of a split is related to the expected time until the next block is found. Bitcoin's 10 minute expected time to mine a block results in a nice tradeoff between the probability of a fork occurring and transaction settlement time.

Blockchain forks happen less than once per month. That's not bad for a network running twenty-four hours a day, every single day of the year. Splits that are two blocks long are expected to occur many years apart. And it's more probable that human civilization will be long gone before we'd ever see an unresolved blockchain split that's 3 or 4 blocks long.

Blockchain splits are another reason why confirmations are important. We mentioned this in the Chapter 5 but convention amongst the Bitcoin network is that six confirmations make a block completely irrevocable. Now you have a lot more information to help you understand why.

Also, recall that coinbase transactions require 100 confirmations before those bitcoin can be spent. In the example above the Cleveland miner and the Tokyo miner don't both get a block reward. Only the Tokyo miner does, but he must wait 99 more confirmations after the London block before he can spend them.

## THE BLOCKCHAIN TRILEMMA

There's a long-standing model in business that states if you want to get something done, you must pick between good, fast, and cheap. You get to pick two, but you'll sacrifice the third. If you want it good and fast, it'll be expensive. If you want it good and cheap, it'll take more time to get. And if you want it fast and cheap, it'll be bad quality.

Blockchain technology faces the same tradeoffs between options. In what's known as The Blockchain Trilemma, a particular blockchain technology gets to choose two out of the following three features: decentralized, scalable, secure.

- *Decentralized* - no central authority or point of control
- *Scalable* - can handle an ever-growing amount of transactions
- *Secure* - can operate without falling victim to attack

From now on, whenever you look at a blockchain technology, know that there must be *some* tradeoff between these three things. It's very likely that a blockchain that claims to have high levels of all three features is a little snake oily. But solutions to solving the trilemma are in the works.

Bitcoin was designed to be decentralized from day one. Its Proof-of-Work makes its security beyond solid. What Bitcoin lacks is large amounts of scalability. It's not built for it.

Currently, Bitcoin can handle a maximum of 7 transactions per second. The Visa payment network can handle about 1,700 transactions per second, but there's not a shred of decentralization in the Visa network. It's completely centralized.

Compared to centralized, intermediary solutions, blockchain technology is inefficient and slow. But lack of scale is the cost of true decentralization and high security.

This lack of scalability is why the core Bitcoin network will always be the standard, settlement base layer, crypto store of value that the entire global decentralized finance system will be built upon. Second layer technologies like the Lightning Network which can handle more transaction traffic and periodically settle on the base Bitcoin layer already exist. More solutions will be built.

## BITCOIN'S ENVIRONMENTAL IMPACT

Satoshi thought long and hard about the fairest way to issue and distribute monetary units from scratch. Proof-of-Work was the answer. The Bitcoin network was designed to use energy and convert it into thermodynamically secure monetary value and validated, verifiable, immutable, and indisputable transaction records.

There's a terrible misconception that Bitcoin is bad for the environment because of mining. This is incorrect. Bitcoin is not an environment killer, and it certainly does not disproportionately hurt the environment more than other technologies. To see why, let's start with the terawatt-hour.

A terawatt-hour is the unit of energy equivalent to outputting one trillion watts of electricity for one hour. The entire world generates approximately 160,000 terawatt-hours of energy each year. Of that production, 50,000 terawatt-hours *are wasted*. The entire Bitcoin network uses about 120 terawatt-hours per year. That's about 0.075% of the world's energy production per year.

Put another way, the 120 terawatt-hours used by Bitcoin each year is equal to 0.24% of the wasted energy. This is an important point. Bitcoin miners generally operate on marginal points of the energy grid, no man's land. You wouldn't see a large mining operation pop-up in the Financial District of Manhattan—the energy there is too expensive, nor would it be politically welcome.

*It doesn't make economic sense to mine where energy is expensive. The numbers don't work out—it'd likely be a losing economic proposition.* Energy is often the most expensive where it is most needed. Miners aren't cutting in line in expensive energy markets, they go where they are welcome and where excess power would otherwise be wasted or stranded.

Miners take marginally valuable or non-valuable energy and use it to mine, perhaps getting the highest marginal utility or power usage of any technology in the world. For about $2 billion in energy costs per year, miners secure over $1 trillion dollars of assets. That's only 20 basis points of security cost a year.

So how much energy *is* 120 terawatt-hours? Well, as a comparison, traditional banking around the world uses 240 terawatt-hours per year. New York City uses about 60 terawatt-hours of energy per year. These two examples alone show that Bitcoin energy usage is *not a big deal*, especially considering what that energy is used for.

But there's always someone in the room who says, "But Bitcoin uses more electricity than a small country!" Well, lots of things in our modern technological world use more energy than a small country. In the United States during every Christmas season more electricity is used on Christmas lights than El Salvador and Ethiopia use during the whole year. But you never hear that story in the news, do you?

As a final note on this topic, I wanted to mention that a lot of Bitcoin energy sources are renewable. Sure, some coal is used in China to mine Bitcoin, but there's an awful lot of hydropower used in China, too. According to the Cambridge Center for Alternative Finance, 39% of Bitcoin mining worldwide is powered by renewable energy sources. This percentage continues to grow.

Mining equipment gets more energy efficient every year and the cost to secure the network naturally goes down as more and more bitcoin are mined, lowering the incentive for the network to burn energy. How much do you think it costs to secure all the world's gold per year? All the U.S. dollars?

All things considered, Bitcoin is actually extremely energy efficient. Frankly, I'd like to see someone come up with a better, more energy efficient way to secure so much money that also has all the benefits and utility that Bitcoin hass.

## THE BYZANTINE GENERALS' PROBLEM

Interestingly, using Proof-of-Work to achieve consensus without a trusted central authority is a solution to a computer science problem from the 1970's called "Byzantine fault tolerance", or the "Byzantine General's Problem." The problem was considered unsolvable until Satoshi came along. It relates to two or more people needing to share information in an unreliable or hostile environment where communications can be lost or tampered with.

Imagine that you and I are each leading an army on opposite sides of a city that we need to attack. Here's the issue: we both need to attack the city at the same time or else we'll be defeated. The city's defenses can easily handle one attacking army. So how can we reach consensus that "OK, this is the time we'll both attack" from opposite sides of the city?

The environment is hostile. Say I send a spy through the city with a message that says, "we attack in two days at 1:37 pm." Once that spy leaves my army, I have no way of knowing whether he'll reach your army. So, should I attack or not? Well, you could send a spy from your army to confirm you received the message, "I got your message, we will attack in two days at 1:37 pm."

But now, you have no idea if I'll receive your message. Should you attack or not? So, I send a confirmation of the confirmation. And then you send a confirmation of the confirmation of the confirmation. This back and forth would continue into infinity. There's no way of knowing that you and I are on the same page due to the communication environment.

Let's complicate it.

What if there were three armies outside of the city that needed to attack at the same time? Let's call the coordinated attack time T. All three armies need to know T. All three armies also need to know that the other armies know T, and that all three armies know that all three armies know T.

Here are the words straight from Satoshi's mouth explaining how Proof-of-Work solves this problem—a first in distributed data processing. Satoshi posted this to the Cryptography Mailing List on November 13th, 2008.

> The proof-of-work chain is a solution to the Byzantine Generals' Problem. I'll try to rephrase it in that context.

A number of Byzantine Generals each have a computer and want to attack the King's wi-fi by brute forcing the password, which they've learned is a certain number of characters in length. Once they stimulate the network to generate a packet, they must crack the password within a limited time to break in and erase the logs, otherwise they will be discovered and get in trouble. They only have enough CPU power to crack it fast enough if a majority of them attack at the same time.

They don't particularly care when the attack will be, just that they all agree. It has been decided that anyone who feels like it will announce a time, and whatever time is heard first will be the official attack time. The problem is that the network is not instantaneous, and if two generals announce different attack times at close to the same time, some may hear one first and others hear the other first.

They use a proof-of-work chain to solve the problem. Once each general receives whatever attack time he hears first, he sets his computer to solve an extremely difficult proof-of-work problem that includes the attack time in its hash. The proof-of-work is so difficult, it's expected to take 10 minutes of them all working at once before one of them finds a solution. Once one of the generals finds a proof-of-work, he broadcasts it to the network, and everyone changes their current proof-of-work computation to include that proof-of-work in the hash they're working on. If anyone was working on a different attack time, they switch to this one, because its proof-of-work chain is now longer.

After two hours, one attack time should be hashed by a chain of 12 proofs-of-work. Every general, just by verifying the difficulty of the proof-of-work chain, can estimate how much parallel CPU power per hour was expended on it and see that it must have required the majority of the computers to produce that much proof-

> of-work in the allotted time. They had to all have seen it because the proof-of-work is proof that they worked on it. If the CPU power exhibited by the proof-of-work chain is sufficient to crack the password, they can safely attack at the agreed time.
>
> The proof-of-work chain is how all the synchronization, distributed database and global view problems you've asked about are solved.

* * *

Ok, that was a long chapter. But there was so much good stuff to cover. We have one final and short chapter in the technical portion of the book. In it we're going to revisit transactions. There's more to them than we've covered thus far. Though I won't cover *all* the details behind transactions, there's a few key points you should understand to up your Bitcoin game.

Before we continue, I want to give you a great Bitcoin joke I found online. Drumroll please...

Why won't the government embrace Bitcoin? They hate the idea of Proof-of-Work!

CHAPTER 13

# TRANSACTIONS

Transactions are the last thing we'll cover in the technical portion of this book. Though you learned a lot in Part 3 about the technology that powers Bitcoin, there's a tiny bit more to cover. The previous chapter on Proof-of-Work was particularly long, so to close out this part of the book this chapter is nice and short.

But don't let the brevity of this chapter fool you. Transactions are important. They're the foundation of Bitcoin. Transactions are what allow a person to transfer value to another person on the network. Everything else in Bitcoin exists so transactions can happen and so the network can keep accurate records of them.

Transactions are much more complicated than they may seem on the surface. However, I don't think the extreme details of transactions are necessary for you to know at this point. Please go ahead and research transactions further if you are inclined; you'll keep yourself busy for a while. For this chapter, we'll keep things as qualitative as possible.

## WHAT'S INCLUDED IN A TRANSACTION

In the real world you might want to send your friend some bitcoin. You pull out your device, open your wallet, and click a few buttons. Maybe you even scan your friend's address in the form of a QR code. In your head, you're thinking "send Steve 5 bitcoin."

As you might have guessed, a lot more is going underneath your wallet's user interface than appears on screen. Here's what a Bitcoin transaction looks like behind the scenes:

```
0100000001813f79011acb80925dfe69b3d
ef355fe914bd1d96a3f5f71bf8303c6a989
c7d1000000006b483045022100ed81ff192
e75a3fd2304004dcadb746fa5e24c5031cc
fcf21320b0277457c98f02207a986d955c6
e0cb35d446a89d3f56100f4d7f67801c319
67743a9c8e10615bed01210349fc4e631e3
624a545de3f89f5d8684c7b8138bd94bdd5
31d2e213bf016b278afeffffff02a135ef0
1000000001976a914bc3b654dca7e56b04d
ca18f2566cdaf02e8d9ada88ac99c398000
00000001976a9141c4bc762dd5423e33216
6702cb75f40df79fea1288ac19430600
```

This real-life transaction was confirmed on the blockchain back in May 2016 in block 410,404. In his great book *Programming Bitcoin*, Jimmy Song walks through the highly technical details of this particular transaction. Check his book out if you're interested in learning how to code Bitcoin. Though I won't go into actual lines of code or the same level of detail here, I will briefly discuss what information is contained inside of a transaction like this one.

A Bitcoin transaction includes three major buckets of data: metadata, inputs, and outputs.

**Metadata** - Important information about the transaction is included. Things such as version number, the size of the transaction, the number of inputs, the number of outputs, and a locktime. Also, the hash of the transaction data is provided, which acts as a transaction ID.

**Inputs** - There can be multiple inputs in a transaction. Each input contains a hash of a previous transaction. This is a reference to previously received bitcoin. Remember, when you initiate a bitcoin transaction you are sending bitcoin somewhere. The transaction specifies *which* bitcoin to send. The input also includes a ScriptSig, which for now you can think of as a signature that proves you own the bitcoin and can spend them. There's also an index supplied for each input, and something called a sequence.

**Outputs** - There can also be multiple outputs per transaction. Outputs describe how much bitcoin is moving and to where. Each output has an amount and ScriptPubKey, which for now you can think of as where the bitcoin is sent.

That above transaction example (and all transactions) contains a lot of information. Let's briefly look into some of the pieces that aren't self-explanatory without getting technical.

Let's start with some of the metadata. The version number of the transaction is included to specify what features the transaction has. This is the first 4 bytes of the transaction.

The last 4 bytes of the transaction is the locktime. Usually, transactions are broadcast to the network, hoping to be added to the blockchain ASAP. But using a locktime, the transaction can be broadcast to the network with the instruction to wait to add it to the blockchain. The wait period can be time-based, or block-number-based. You can basically say to the network, "wait one year from today to add this to the blockchain" or "wait until block (any future block number) to add this to the blockchain."

As for transaction inputs, each one includes a sequence number. This is a 4-byte number that was used for an old feature that's now disabled. It's only used now for disabling locktime. However, those bytes remain a part of each transaction input. Perhaps in the future, a cool new feature will use them. They won't go to waste.

Next are the ScriptSig and ScriptPubKey. The ScriptSig for each input and the ScriptPubKey for each output are little, tiny computer programs within the transaction. The language they're written in, called Script, is not Turing Complete and does not contain loops or any control flow operations besides conditional control flow. That means that these mini programs are totally predictable, aren't complex, and are completely immune to a variety of attacks. The language is purposefully limited to eliminate vulnerabilities.

The ScriptSig piece of an input is called the unlocking script. This script contains the conditions necessary to unlock the bitcoin and make them spendable in the current transaction. The ScriptSig also contains a digital signature and a public key.

The ScriptPubKey piece of each output is called the locking script. It specifies the conditions necessary to spend the output in a future transaction. *The outputs of one transaction can be spent as inputs in another transaction.* The destination address for the bitcoin is also included here.

It's at this point where I could spend an entire chapter on ScriptSig and ScriptPubKey. But I won't. You can dig into those on your own if you're interested. These transaction inputs and outputs may seem confusing. So, let's look at how the Bitcoin system accounts for value. That's much more important than understanding exactly what the ScriptSig and ScriptPubKey mini programs do.

## ACCOUNTING FOR VALUE IN BITCOIN

Your bank and credit card companies use an account-based system to track *how much money each account has.* Every statement you receive from a financial institution has an Account Number somewhere on it. And when you make a payment, the institution checks your account balance to see if you have the funds before processing that payment.

Bitcoin doesn't work this way because Bitcoin does not have accounts like you're used to. Instead, it uses a unique type of accounting developed by some well-known cypherpunks to track value known as the UTXO model.

### Unspent Transaction Outputs (UTXOs)

UTXO stands for "*unspent transaction output*" and is a reference to specific bitcoin. UTXOs are the references to spendable outputs, the bitcoin in circulation that are available to spend. At all times Bitcoin full nodes keep track of all the UTXOs out there. This list is called the *UTXO set.* Keeping track of the UTXOs helps nodes validate transactions quickly.

The UTXO set grows and shrinks in size over time, and every single transaction changes the UTXO set. As of the time of this writing, the total amount of UTXOs is approximately 75,000,000 and the size of the serialized UTXO set is about 4.2 GB.

This may seem a little complicated. But a quick example will show you it's not. We'll use two people to get the concept across, Sarah and Jeff. Let's say that Sarah is a miner and just solved a Proof-of-Work for a block. Her block reward is 6.25 bitcoin. A new UTXO is created for those bitcoin. Let's call it UTXO-Sarah-A.

Next, Sarah wants to pay Jeff 2 bitcoin, so she creates a transaction. Her transaction uses its ScriptSig to unlock UTXO-Sarah-A and all the 6.25 bitcoin within it, making them available to spend. The transaction creates a new UTXO for Jeff that contains 2 bitcoin and sends it to his bitcoin address. Let's call this one UTXO-Jeff-A.

But the transaction also sends the remainder, 4.25 bitcoin back to Sarah. This is called *change* in Bitcoin. It's just like when you use a $10 bill to buy a $6 latte. You get $4 back in change. A new UTXO for Sarah is created, UTXO-Sarah-B, that contains 4.25 bitcoin. At this point, UTXO-Sarah-A is destroyed. We're done with Sarah's UTXOs. Let's focus on Jeff again.

Maybe Jeff already owns 8 bitcoin referenced in a previous, but still existing, UTXO. Let's call that UTXO-Jeff-Z. With Sarah's 2 bitcoin payment, Jeff now owns 10 bitcoin in total, spread across two different UTXOs.

Now assume that for some reason, Jeff owes his friend Nancy 9 bitcoin. To pay her Jeff creates a transaction which unlocks the 8 bitcoin in UTXO-Jeff-Z *and* the 2 bitcoin in UTXO-Jeff-A.

The transaction sends 9 bitcoin to Nancy's address (UTXO-Nancy-A) and sends Jeff 1 bitcoin in change (UTXO-Jeff-B). Both UTXO-Jeff-A and UTXO-Jeff-Z are destroyed.

See, UTXOs are simple.

## IMPORTANT UTXO POINTS

Here are some important points regarding UTXOs that are worth knowing and remembering:

- UTXOs prevent double-spending. If a transaction tries to spend an output that's not in the UTXO set it's proof that it's a nonexistent output that's trying to be used as a double-spend. It would immediately be considered invalid and be thrown out by the network.

- UTXOs are always all or nothing. They are always completely consumed during a transaction. You can't partially spend a UTXO. As seen in the above example, you may receive a new UTXO as change though.

- It's impossible for two transactions to affect the same UTXO.

- UTXOs create more privacy than an account-based system. If new addresses are used for each transaction, it's very difficult to link specific coins to specific people.

- When looking in your wallet you don't have a "balance" of bitcoin like you have a balance of dollars in your bank account. What you do have is a bunch of UTXOs (each controlled by a private key) that may come from hundreds or thousands of individual transactions spread out across blocks all over the blockchain. Your wallet detects the UTXOs you own and sums them up for you into a nice total.

- Your wallet automatically handles selecting the most efficient UTXOs to use as inputs in your current transaction. You don't have to deal with that.

* * *

Well, that's enough on transactions. There's a lot more I could have dived into, but I think you've worked hard enough in Part 3 of the book.

If you decided to dig deeper into transactions, perhaps investigate the details of what ScriptSig and ScriptPubKey do or look into Multi-signature wallets (MultiSig) that allow for multiple private keys (from different people) to control specific funds. If you want to open an entirely different can of worms, check out Segregated Witness (SegWit).

Great work on all the technical stuff in Part 3. It's not easy but if you do understand the book up to this point **you can safely and confidently say that you know more about Bitcoin now than 99% of people (actually, it's probably closer to 99.9% or more).**

Don't take that as a joke. I'm serious. Bitcoin is difficult and there are a lot of moving parts. *You really did learn an enormous amount so far.* If you pick someone randomly off the street, even a crowded Manhattan street, I'd bet real bitcoin that you know more about Bitcoin than they do. That's worth celebrating—go buy some bitcoin for yourself!

Which brings us to Part 4—Investing in Bitcoin.

It's time to learn how to invest in and own Bitcoin so you can dramatically change your financial future. Onwards to the final portion of the book…

# Part 4

## *Investing in Bitcoin*

*Rules for people who want to take this seriously. Buying and selling bitcoin. Storing keys. A wallet approach. All-important seed backups and wallet safekeeping. Investment strategy. Telling financial advisors to keep their hands off your bitcoin. What to do with your bitcoin when you die.*

CHAPTER 14

# 15 RULES FOR INVESTING IN BITCOIN

We'll start off the final part of *The Bitcoin Shortcut* with some rules for investing in Bitcoin. I suggest that you read these now and come back to them again after you've finished the book. They're here to make your Bitcoin investment journey as smooth and stress-free as possible, so it wouldn't hurt to revisit them occasionally.

You'll be converting your home fiat currency into BTC. But it's more than just that. You'll be buying into the world's future standard, base-layer store of value—the hardest, most advanced form of money in the galaxy—Bitcoin. It's serious business but at the same time it's ultra-exciting.

These 15 rules aren't in any particular order but they're all important. Ignore them at your own peril. Some of these rules contain ideas that we haven't covered yet. Don't worry, we'll get to them in this part of the book.

**Understand Bitcoin.**

Don't invest a single cent into Bitcoin until you're comfortable with your level of Bitcoin knowledge.

Convert your fiat into Bitcoin with conviction, not on a whim. If you don't understand what you're investing in, then what you're doing is purely speculation. There's no sympathy for the uneducated or undereducated in Bitcoin world.

**Take responsibility.**

Investing in Bitcoin and gaining the financial self-sovereignty that comes with it, requires much more responsibility than any traditional investment ever could. It's your obligation to take on this financial and intellectual responsibility yourself; it can't be outsourced. There's no alibi for the irresponsible. If you can't be responsible (and accountable) perhaps Bitcoin isn't for you.

**Take storage and custody seriously.**

Investing in Bitcoin doesn't end once the purchase is made. Understanding what custody and storage options are available to you and which ones best fit your needs is critical. Self-custody and private key storage are no joke.

**Always control your keys.**

When you purchase bitcoin make sure that you receive your private keys. Some services allow you to buy bitcoin but do not give you control of the keys. Avoid these services like the plague. Not your keys, not your bitcoin. Be careful where you buy bitcoin.

**Buy bitcoin directly.**

Buy bitcoin on an exchange or through your wallet if possible. Don't get sucked into buying any available Bitcoin ETFs (exchange-traded funds), mutual funds, or any other Wall Street created investment vehicle tied to Bitcoin performance. Avoid the unnecessary management fees, lack of private key ownership, and any "Investment Advisory Boards" shuffling papers around pretending to be Bitcoin experts.

Nothing a financial advisor or Wall Street offers you is better than buying bitcoin yourself. You don't need any sort of product or fund; you just need bitcoin.

**Trust yourself and think independently.**

Don't ask someone when you should buy or how much. Don't ask someone when you should sell. Don't offload critical decisions and be shackled to someone's opinion. Trust yourself and make your own decisions. The way to acquire good judgement and confidence with Bitcoin is through education and action. Keep learning and get started on your Bitcoin investment journey.

**View Bitcoin as a non-overlapping asset.**

Stephen Jay Gould, evolutionary biologist, said that science and religion are *non-overlapping magisteria*, meaning that each represents different areas of inquiry—the two domains do not overlap. In this spirit, view Bitcoin as what I call a "non-overlapping asset." Bitcoin is a unique beast like no other asset in the history of Earth itself. Bitcoin and *all other investment assets* that Wall Street offers access to are in two separate domains that do not overlap.

Bitcoin should not be viewed as a "part" of your portfolio of investments. It's a store of value that *is* its own portfolio. Don't group Bitcoin in with all your other investments. From now on, you have a traditional portfolio of investments, and you have Bitcoin. *Keep them separate in your mind and treat them like two separate entities.* These kids don't play in the same playground. Wall Street will disagree.

**Never pay financial advisors any fees on your bitcoin.**

Under no circumstance should you ever pay a management fee, performance-based fee, commission, or any other type of fee to a financial advisor for your Bitcoin investment. When you truly understand Bitcoin, this rule becomes painfully obvious. Again, Wall Street will disagree. Stay vigilant.

**Don't day trade.**

Don't try to time the market and day trade BTC. You'll lose. Instead buy bitcoin, ignore the daily volatility, and hold on for dear life (*HODL*). Real Bitcoin investors who embrace the technology and ethos don't day trade. It's not even on their minds. They want to HODL as long as possible. Your goal should be to acquire as much bitcoin as you can and don't sell a single satoshi for decades. Patience. *(HODL is a widely used term in the Bitcoin community. It's pronounced "hawdle", like bottle with soft d's, boddle…hoddle, HODL.)*

**Dollar-cost average.**

It's the only investment strategy you need for Bitcoin. Have the discipline to follow it. Ignore all other strategies. Ignore fancy technical analysis indicators (bullshit). Avoid complication. Embrace simplicity.

**Stay focused.**

The Bitcoin game has just started. We are unbelievably early. It has been said that: "We're on the second pitch of the first inning in terms of where Bitcoin is today." Bitcoin has a *very long* way to go. Stay focused and think in terms of the 10-, 15-, 20-, and 50-year time horizons. Think about your kids and grandkids. Buy bitcoin for them. Ignore criticism from naysayers who don't understand Bitcoin. Don't be distracted by short-term news. Put up your defenses against media fear, uncertainty, and doubt.

**Stay aware of legal and tax considerations.**

Things can change rapidly as the world adjusts to Bitcoin. Keep tabs on developments. Worst case, you lose your bitcoin in a boating accident.

**Avoid the risk of ruin.**

There's always a chance you can lose a lot of money through market forces or mistakes. You must stay in the fight. Don't allow emotion or logic bugs to take over. You must avoid going broke. Don't put any fiat into Bitcoin that you can't afford to be without for many years. It will be invested for the long term and should be worth more, but there's always a chance that you'll lose it all. That's your responsibility. Don't do anything stupid or illegal. Avoiding an orange jumpsuit is just as important as not going broke.

**Have congruency and regularly evaluate your stances.**

Hone your economic, political, and philosophical viewpoints. Be able to articulate them. Think and act in agreement and in harmony with the ethos of Bitcoin. It takes real courage to do this, sometimes at great risk. You fall on an economic and philosophical spectrum, somewhere between total conformist sheep and fully autonomous self-sovereign individual. Identify where you stand and how Bitcoin plays a role in your new life. Have skin in the game. Know whether you'll give up and give in at the first sign of discomfort, or if you're the type of person to switch jobs, move to new jurisdiction, or even renounce your citizenship to stay true to your beliefs and opinions.

**Be skeptical of other cryptocurrencies.**

You don't have to avoid or dismiss other cryptocurrencies. I'm not saying that. Bitcoin and other cryptocurrencies surely will coexist in the future, no doubt. Just be careful before investing in other coins. Ask yourself, what does the cryptocurrency you want to buy *do* that Bitcoin does not? What tradeoff does the cryptocurrency make to accomplish the utility it claims to provide? Be careful and do your due diligence before investing in anything in the crypto space. It's very unlikely that another cryptocurrency replaces Bitcoin's role as the base-layer store of value in our future world of fully decentralized banking, finance, and investments. Make sure you know *exactly* how your potential investment differentiates from Bitcoin.

* * *

CHAPTER 15

# BUYING AND SELLING BITCOIN

It's never been easier to buy and sell Bitcoin. Although it may seem intimidating, especially if you've never done any traditional investing before, I promise you that you can handle it.

In this short chapter we'll discuss how you can get your hands on some actual bitcoin. There's not a whole lot you need to know, so let's blast through it.

## CRYPTOCURRENCY EXCHANGES

The most popular way to purchase Bitcoin is through a cryptocurrency exchange. An exchange is a marketplace that allow users to buy and sell different cryptocurrencies. Through an exchange you can convert your fiat money into Bitcoin or any other listed cryptocurrency. Exchanges also offer high liquidity, low transaction fees, and real-time market data.

Typical methods of funding your exchange account include bank transfer (ACH), wire transfer, debit card, and credit card. This varies by exchange and jurisdiction. Direct deposit of Bitcoin or other cryptocurrencies is a standard capability, too. With a crypto deposit the exchange will provide you with a bitcoin address to send the deposit to.

Since the crypto world is evolving quickly, different exchanges have different processes, perks, rules, and standards. But the one thing they all have in common is that they allow you to buy and sell. Regulations and oversight of exchanges vary by jurisdiction, too. If you're in the United States and using an exchange based in the US, you can feel safe that the company is not going to scam you.

Two exchanges that I recommend are Gemini and Coinbase. This is not a paid endorsement for either. I use them both and am just a fan. Their onboarding processes are simple and user friendly. There are pros and cons to each of these exchanges but overall, I think using one of them to start is your best bet. But of course, do your own research and search online for an exchange that best meets your needs.

## OTHER WAYS TO BUY

It's worth quickly mentioning that there are other ways to buy Bitcoin that don't directly use an exchange. Many dedicated wallet apps and third-party financial apps like PayPal allow you to purchase and sell Bitcoin.

You can also use more direct person-to-person services like LocalBitcoins.com or use one of the thousands of dedicated Bitcoin ATMs around the world. A quick search online will help you locate one near you.

Disadvantages of these purchasing options may include lower liquidity, higher fees, and no access to private keys.

*Always do your homework before making any purchase no matter which method of acquisition you choose. You're responsible. Remember that in Bitcoin world mistakes can't be reversed.*

## MINIMUM PURCHASES

I remember when Bitcoin was around $50,000 per coin someone said to me that they couldn't afford to invest because it was so expensive. I guess it's not obvious to everyone, but you *do not have to buy a whole bitcoin to invest.* If you want to buy a whole coin go right ahead, but if you only want to invest $1,000 or $100, or even $10, you can.

So lucky you, there's no excuse not to begin accumulating bitcoin! You don't need to be rich to start. The only real limitation will be the minimum purchase order required by the exchange or service that you use to buy bitcoin. This is usually around $5 per purchase, but it will vary. One app I have used has a minimum purchase of only $1.

So, you see, *anyone* can start building wealth by investing in Bitcoin. As one of my old mentors used to say, "You can't afford *not* to do it."

## KYC / AML

Depending on what exchange or service you use and where you live you may be subject to KYC and AML during onboarding. KYC stands for Know Your Customer and AML stands for Anti-Money Laundering. These are regulatory policies that are in place to make sure you are who you say you are and that you don't do any funny business.

Typically, you'll fill out the usual personal information when signing up for an account such as name, address, and so on. Then you'll likely have to include your Social Security Number and an upload of a government issued ID.

It's easy and fast. And it's no big deal.

## ORDER TYPES

When trading on an exchange you can use different order types to buy or sell in the marketplace. The most common of these order types are called market, limit, and stop orders. Each of these types of order serves their own unique purpose. Here's a quick summary of each.

**Market Order** - This is an order to buy or sell, immediately, at the best available price. This type of order practically ensures the order will be executed (completed) but it gives no guarantees of the price. You may overpay or underpay compared to the price you see on your screen. If the actual cryptocurrency you are trading has high liquidity, the price will be close to the last price you see on the screen. Some exchanges will warn you if there is too much slippage while preparing your trade, meaning the price moved a lot before you could execute the order. This gives you an opportunity to adjust your trade details if you desire.

**Limit Order** - This is an order type where you can specify the exact price to buy or sell. There's no guarantee the order will be executed, but if it does, you're guaranteed to get your specified price or better.

**Stop Order** - This is an order to buy or sell at the market price but only after the price of the asset has traded through the specified stop price. For example, pretend Bitcoin is trading at $50,000. You think it's going to take off but aren't sure. You want to wait and see, but you don't want to wait too long and miss too much upside. You set a stop price at $51,500. If Bitcoin's price goes past that, your order immediately becomes a market order and executes at the next price. But if Bitcoin never reaches the stop price, the order will never become live. Bitcoin may go up to $51,499.99 and your order won't execute.

Or perhaps you want to limit losses. Say you're comfortable with a 10% downward swing, but after that you want to sell. In that case you'd set a stop price at $45,000. Bitcoin could trade anywhere in between $45,000 and $50,000 and your trade wouldn't execute. But once it hits that lower bound of $45,000 your stop order becomes a market order and will execute the sale at the next available price.

All these order types are useful in certain situations but most of the time you'll probably just use market orders. There are more order types available such as stop-limit, market-if-touched, and limit-if-touched, but you don't need to complicate your life so much. Keep it simple.

## TRANSACTION FEES

Exchanges and other platforms need to make a living. One way they do this is to add on a very small transaction fee to your buy and sell orders. These transaction fees vary by exchange and service; there's no standard. Some factors that can affect the level of the fee include, but are not limited to, account size, trade size, order type, the cryptocurrency traded, web or mobile, special platforms, and so on.

It's totally normal to pay a small fee for each trade either in percentage terms or dollar terms. Just make sure the fee isn't too much. If it is, find a better solution. Overall, a transaction fee should not be a huge percentage of the trade. For example, you shouldn't be paying $5 to purchase $50 of Bitcoin, that's too much. That'd be a 10% transaction fee. No one should be crazy or stupid enough to pay a 10% transaction fee.

## OPEN 24/7/365

The stock market is only open for trading from 9:30am to 4:00pm on weekdays. It's closed on weekends and holidays, too. It's not even open 20% of the entire 168 hours in the week. Bank hours are better, but not by much.

Bitcoin is open for business 24 hours a day, 7 days a week, 365 days a year. Arbitrary time frames and holidays can't stop the exchange of value in Bitcoinland. It's always there for you, like a great friend.

* * *

Rest assured, once you choose an exchange and get your account set up, buying some bitcoin will be as easy as clicking a few buttons. The Bitcoin ecosystem is really doing a great job of making it easy for new users.

While you could start investing today, I recommend that you finish the book first. The final two chapters of *The Bitcoin Shortcut* cover how to store bitcoin and some Bitcoin investment strategy. Let's keep going.

## CHAPTER 16

# STORING BITCOIN

Probably the most neglected aspect of Bitcoin investing and ownership is storage. This is especially true for people just getting involved who often overlook it completely.

Storage isn't sexy, but it's damn important. Mistakes executing your storage strategy can result in some rude awakenings down the road. In the worst-case scenario, the total loss of all your bitcoin investment is a possibility.

There are many ways to implement your storage strategy—from totally passive and indifferent, to fully active and meticulous. For certain, your storage choices will be heavily influenced by how serious you are about Bitcoin and how much responsibility you want to take on.

Recall that the monetary units of Bitcoin live and stay on the Bitcoin blockchain. What you own and need to protect when investing in Bitcoin are the private keys that provide access to your bitcoin. Whenever you hear anyone talking about storing bitcoin, whether they realize it or not, they're talking about storing private keys.

How you choose to store your private keys is a big deal. Make sure you understand this chapter and take the time to think through what storage strategy works best for you.

There are no rules. You can do this however you want—it's your money.

Let's start with the basic idea of *custody*.

## CUSTODY

You can't beat the Oxford definition of custody. It defines it as, "the protective care or guardianship of someone or something." Custody also relates to the ability to *move* assets at will.

You have two options for the custody of your private keys: self-custody or third-party custody. Self-custody means that *you* store and protect your private keys. Third-party custody means that *someone else* stores and protects your private keys. You can mix and match between these two options.

All bitcoin private keys are stored in wallets. Every single type of wallet in existence can fit into one of two categories: noncustodial and custodial. Noncustodial wallets give users total control over their private keys while custodial wallets (and the keys they store) are controlled by trusted third parties.

The most common custodial wallets are cryptocurrency exchanges. The exchange will keep accounting records of how much bitcoin you own, but the exchange will control the keys. This raises the question: *are they even your keys?*

How exactly an exchange manages the safety and security of these keys and accounting records varies. There are risks associated with exchange custody which we'll discuss soon.

Let's make this idea of custody very clear. When you self-custody, you *control* the keys (your bitcoin). When you rely on third-party custody you *do not control* the keys (not your bitcoin), though through their user interface it may really look and seem like you do.

Custody means a lot. Choose custody options wisely. By the end of this chapter, you'll have a better idea of what you want to do.

## WALLET TYPES

Wallets are considered hot or cold, referencing whether they are connected to the internet or not. A *hot wallet* is connected to the internet most or all of the time, meaning the private keys are quickly and readily available for use in transactions. A *cold wallet* is not connected to the internet, the keys are stored completely offline and are not quickly accessible. Hot wallets are vulnerable to online attacks but are much easier to use. Cold wallets are less easy to use but are designed to be immune from attacks.

When we take into consideration hot and cold, as well as custodial and noncustodial, we have four combinations: hot custodial wallet, hot noncustodial wallet, cold custodial wallet, and cold noncustodial wallet.

Going a little deeper, you'll see that you also have many variations of wallets. They include web-based, desktop, mobile, hardware, and paper.

**Web** - These wallets are stored on servers owned by third-party entities and accessed with a typical web browser. The keys are not usually controlled by the user. Ease of use and ease of access are benefits but the reliance on third parties (which you are forced to trust) is a pain point many aren't willing to deal with.

**Desktop** - This type of wallet software runs on a Mac, Linux, or Windows computer and private keys are saved directly on the hard drive. Private keys are controlled by the user and the wallet can be used offline. But any software sitting on a computer has potential security vulnerabilities.

**Mobile** - A mobile wallet runs on a smartphone. These are the most popular form of wallet used today. They're very user friendly and many offer users the opportunity to control their own keys. There are a few things to keep in mind regarding mobile wallets. First, the security might not be fantastic. For example, if neither your phone nor wallet has a passcode, anyone will be able to access your funds if you lose the phone.

Also, you must make sure that the wallet you download is actually a wallet app that stores private keys *on the device.* It's very possible that the wallet you download is just a web-based wallet from a third-party that provides access via the app. Looks like a mobile wallet, feels like mobile wallet, but isn't a mobile wallet. Big difference.

**Hardware** - A hardware wallet is a secure, dedicated device engineered solely for the purpose of storing bitcoin and other cryptocurrencies. Most look like USB drives but are way more advanced. Special-purpose software runs on the device to securely store private keys. Hardware wallets are cold storage that only connect to your computer and the internet at the time they need to execute a transaction. The actual digital signing of these transactions is done "on device" and for some brands, physically pushing some buttons is also required to perform a transaction. Hardware wallets are safe against attacks that can potentially affect hot wallets. For storing large amounts of money, hardware wallets are the best option.

**Paper** - Qualifying as cold storage, paper wallets are private keys printed out on paper. Non-paper material also works fine, as long as the private keys can be printed, written out, etched, or engraved. Once the keys are on the material, the wallet is stored somewhere so it is safe from physical harm.

It's worth looking into these wallet variations to see what best suits you. In the section Wallet Strategy later in this chapter, I give you my opinion of what types of wallets to use and how to keep them safe from harm.

## BITCOIN CUSTODY AND WALLET RISKS

This is not an exhaustive list of the risks associated with custody and wallets, but it does include important things to think about when deciding on your bitcoin storage strategy.

**Counterparty Risk** - Any method of custody that is not self-custody introduces counterparty risk to your overall bitcoin investment plan. There may be many reasons why a third-party can't or won't pay you when you ask for your funds. Generally, they'll be able to pay you on demand. But what happens if they need deal with financial problems, regulatory issues, banking problems, or technical issues at the time you want to access your money? Do your research to see how the service you plan to use handles potential counterparty risks.

**Exchange Hacks** - While the Bitcoin network has no third parties, crypto exchanges *are* third-party on-ramps to the network. Third-party systems always introduce more layers of security vulnerability. When you use an exchange to store your private keys, you're relying on their security measures and are helpless if an attack occurs. Always be cognizant of the exchange's security procedures.

**Legal and Political Risks** - Private keys stored by third parties *always* have the potential to be seized due to legal or political reasons.

**Destruction Risk** - Whether the keys are stored in a hot or cold wallet, in self-custody or with a third-party, there's always risk that the keys could be destroyed. This risk must be seriously analyzed and mitigated as much as possible. If you have control of your keys, this mitigation includes backing up your seed phrase.

**Poisoned Device** - Buying hardware wallets from secondary sources, not directly from the manufacturer, always comes with risk. I wouldn't do it. Devices sold on secondary markets can have preprogrammed seed phrases that the seller secretly knows. Later, they can wipe out your funds. Only buy hardware wallets directly from the manufacturer's website. And never use a wallet pre-loaded on a computer or smartphone. If you buy a device on the secondary market, wipe it clean first before installing a new software wallet.

**Imposter Apps** - Always download software wallets directly from the company's website or through their direct download links. Apps that impersonate the real wallet app company exist. Sometimes the App Store and Google Play Store can't get rid of them fast enough and people have been known to have 7-figures worth of crypto stolen from them by the criminals behind these malicious apps.

**Phishing** - A phishing scam is one that targets consumers through emails that appear to be from a trusted source. Say that you use a wallet from Company XYZ. Months from now, you get an email that appears to come from Company XYZ stating something intriguing that gets you to click a link or a button. The goal is to gain access to your device or get a hold of important information about you or your account.

Always check the sender's email address. If it's not directly from Company XYZ, delete the email and don't click anything inside the email. In today's fast-paced world where attention spans have dropped to one or two seconds, it's very easy for even the smartest people to get fooled by a sophisticated phishing attack.

**$5 Wrench Attacks** - This is a term used in the Bitcoin community to describe any physical attack that results in the theft of your bitcoin.

If somebody knows that you own a lot of bitcoin it's not difficult for them to attack or threaten you to get their hands on it. A bop on the head with a simple $5 wrench is all it takes to steal your smartphone, computer, or hardware wallet.

The best you can do to avoid $5 wrench attacks is keep private about your bitcoin investments. Don't tell anybody how much you invested, when you invested, or how much your bitcoin is currently worth. It's none of their business and only makes you more vulnerable.

Don't walk around with your hardware wallet around your neck like a piece of jewelry. Don't put it on your keyring. Don't post screenshots of any useful information regarding your bitcoin to social media. Improve your home security, learn some self-defense techniques, and even set up dummy wallets for situations when you're in a total bind.

Be discreet and know that physical attacks for bitcoin are a real thing. Hopefully you never have to deal with a $5 wrench attack, but you should be prepared for it. If you can't run or hide and must fight—fight to win.

## PRIVATE KEY BACKUPS

If you've decided on self-custody, the moment you start using a brand-new wallet your 12- to 24-word seed phrase will be displayed. It's important to immediately back it up. Do not take a screenshot or picture. Instead, write the words down on a physical piece of paper.

Do this somewhere quiet and private. You need to concentrate. If you mess up the initial backup of your seed phrase, you're screwed. Remember, there is no such thing as a "forgot password" button to help you later. *You must immediately take responsibility for your seed phrase.*

This quick paper backup is helpful and important, but you need a much more permanent and secure form of backup for your seed phrase. The paper you wrote your seed phrase on should be destroyed after you employ a more serious backup method.

Though many methods exist, here are five ways of backing up your seed phrase. Choose whichever one (or more) that fit your needs. Remember there are no rules here, the Bitcoin police aren't going to come and get you. Do what makes *you* the most comfortable.

**Mental Memorization** - Memorizing your seed phrase provides the ultimate form of protection for your bitcoin. No one can extract those 12 to 24 words out of your head unless you allow them. Obviously, this backup form has a lot of risk. If you forget your seed phrase you're out of luck. If you lose the wallet or it gets destroyed, you'll never be able to access those funds again since you can't restore your wallet without the seed phrase.

Are you willing to risk a total loss of your funds for 100% protection of your seed phrase?

Keep in mind, accidents do happen. What if you're in a car accident and for some reason lose your short-term memory? Funds gone. Or what if you're incapacitated or killed and loved ones have no way to access your seed phrase? Funds gone.

Only you can make the judgement if this should be your *only* method of backing up your seed phrase. As a secondary or tertiary form of backup, everyone should do it. Memorizing a list of 12 to 24 words is not difficult at all. But as a one and only form of backup? That's up to you.

As an interesting note, here's a way this backup method can be (and has been) utilized. A person wants to leave their country for better pastures, but are afraid of search, seizure, or capture. First, that person mentally backs up their seed phrase and destroys all physical copies of it. Then they perform their escape. Once safely in their new land they quickly procure a new wallet and restore their keys using their memorized seed phrase. After this restoration they more permanently back up their seed phrase and go on their jolly way.

The monetary value has safely and secretly moved across jurisdictions without ever being in the hands of the owner. There's no wallet or account to censor or seize. It's a mental transfer of value across time and space that's completely unstoppable—try that with fiat or commodities.

**Digital** - At the end of the day digital backups are just files that you must access via a computer or smartphone. Don't write down your seed phrase and save it as a Word file titled "Seed Phrase" and conspicuously keep in on your Desktop. Be creative.

Perhaps coming up with an algorithm that you can use to decrypt a message containing your seed phrase would work. Or maybe write a short story, with each paragraph starting with a seed word. Think of interesting ways to hide the information within the file. Then password protect the file and store it offline and off-device on a USB drive or external hard drive.

Remember, digital files saved on a device that is or will be connected to the internet are always at risk. Plus, computers, phones, hard drives, USBs, and similar devices can always be lost, seized, or destroyed. Keep this in mind and plan accordingly.

**Steganography** - This is the art and science of concealing secret information within computer files. As a great example, imagine you have a folder of pictures from your last vacation. They aren't particularly interesting, and most people would breeze right by them if they were looking through your computer for important information. Yet, one of those images could have your seed phrase secretly embedded in the image. This embedding is done in a such way that even people who are highly proficient with computers detect the information. Steganography is a highly specialized topic and a wonderful rabbit hole to explore.

**Paper** - You could write your seed phrase on a simple piece of paper. That works just fine if you protect it from damage and theft. Laminating it is also an option for added protection against liquids. Make sure to write the words clearly. Better yet type them. Perhaps think of a way to encode them.

**Metal plates** - This is a great primary seed backup method. You can engrave or stamp your seed phrase onto metal plates. This method has the advantage that heat, chemicals, and blunt trauma can't destroy the backup. You can purchase many different brands online. Just search for "metal bitcoin backup", "metal seed plate" or similar phrases. You'll find some. They're worth every penny.

### *Seed Splitting*

Seed splitting is the idea of taking your seed phrase, splitting it into two groups of words, and backing up each group separately. The two backups are then stored in different physical locations.

There are pros and cons to seed splitting. I suggest you look into it more and decide for yourself if it's a good idea for you. I just wanted to expose you to the idea. There's no way I could recommend or not recommend seed splitting to you without knowing your exact situation and having a long in-person conversation with you.

## WALLET SAFETY AND STORAGE

One popular question newcomers have is: "If my wallet gets destroyed do I lose my money?" Great question. The answer is no, you won't lose all your money.

If your wallet ever gets destroyed, you can simply get a new wallet and restore your private keys from your seed phrase. Don't worry. But of course, if your wallet and only seed backup get destroyed at the same time, yes, you'll lose access to your money.

If you're using a hot wallet on your smartphone the physical safety of that wallet is only as good as the way you treat and handle your phone. I'm not here to lecture you on how to treat your smartphone. Instead, what I want to discuss is how to keep a cold wallet safe from harm.

The biggest mistake you can make is to treat a hardware wallet like some junk USB drive that you got for free at a lame work conference. You don't want to throw your cold wallet into a kitchen drawer and forget it's even there.

With self-custody of bitcoin, you *are your own bank vault*. Treat your cold wallet with respect.

You do have some tradeoffs you must consider for the physical safety of your cold wallet, though. They are security, fire, water, crushing, and portability.

Let's start with security. The most secure storage for your cold wallet would be one of those multiple thousand-dollar, heavy metal security safes that are bolted to the ground in your basement. There's no chance a robber in your house is getting that out of there. Yet, it's obvious that a safe like that has valuables inside it, so you never know what they'll do.

On the other hand, a small jewelry safe that you'd perhaps stick in your closet is secure, too. But those are very easy to steal and break into later. If you decide to lock up your cold wallet, how far do you want to go?

As for fire, what matters is keeping your cold wallet in a place or container that's fireproof. Different safes, cases, and bags are graded differently for fire and heat, so make sure you confirm this information before purchasing one. Or, placing your wallet in a case and burying it somewhere is an option. A metal safe that's a few feet underground in your backyard is protected from fire.

It's important to worry about water, too. Pipes break, leaks and floods happen, and even if your cold wallet is in a fireproof case and your house burns down, many of those cases aren't waterproof. Maybe the firemen save your house but destroy your cold wallet. It's worth taking the time to find protection that's fireproof *and* waterproof.

Crushing is also something to keep in mind. There's a surprising number of scenarios where your device can get crushed if you start thinking about them. It's worth considering. Look for cases and safes that are resistant to physical crushing.

Finally, quick portability is a consideration. Would you be OK if you had to leave your house in an instant and your cold wallet was stuck in your safe in the basement? What if you couldn't return for days, weeks, or months? Or would you prefer your cold wallet to be in a grab bag that you can quickly take with you if needed? Think about it.

Between these five variables there are a lot of combinations and tradeoffs. Only you can decide what's best for your bitcoin.

I'd also like to briefly mention safe deposit boxes. I do not recommend this option for a cold wallet because you're introducing a third party. Your custody turns into a strange blend of self-custody and third-party custody that's has its own problems.

If you do choose this route, research your country's, state's, and city's laws. Your safe deposit box might be searchable or seizable. Also, access to your box may be limited by the institution you're renting it from. What if you need to access your cold wallet immediately and your bank is closed?

How you decide to protect your cold wallet is nobody's business. Keep your methods private and don't introduce unnecessary security risks.

## WALLET STRATEGY

If you decide to utilize third-party custody in your bitcoin investment strategy, then the main decision is which exchange to use. Different exchanges have different procedures and security protocols. They may or may not have cold storage insurance to protect your funds from specific storage risks. You should confirm the details of the specific exchange before committing to it.

With this custody option though, the gist is that you'll just treat your bitcoin exchange account like any other investment account. You'll deposit fiat money into it, buy bitcoin, and let it sit there. You'll most likely be able to transfer your bitcoin from the exchange to your own wallet in the future if you desire, but there may be hold periods. During the time your Bitcoin investment is sitting in an exchange account, the private keys are *theirs* and you are exposed to certain risks we discussed earlier.

I can tell you this:

**All serious Bitcoin investors and supporters choose self-custody.**

It goes with the territory and fits with the ethos. To take advantage of the total financial control of your funds and approach the goal of total self-sovereignty, you need to utilize self-custody. And you should primarily use cold storage.

Here's my recommendation. If you do choose to be serious about Bitcoin, start off by getting yourself a hardware wallet, a metal backup solution for your seed phrase, and the necessary physical protection for both.

A good hardware wallet will cost roughly $50 to $200. I recommend COLDCARD hardware wallets, but new great brands will certainly pop up. As for metal backup plates, Blockplate, CryptoTag, and SEEDPLATE are all fantastic choices. These range in price from about $50 to $350.

When it comes to physical protection of your cold wallet you can find all types of durable safes, fireproof cases and bags, waterproof bags and tubes, and crushproof cases and tubes online with minimal searching. You have a ton of options for relatively inexpensive prices. I don't recommend being too cheap about it though—this is your bank vault you're protecting.

As for a strategy for *using* wallets, here's a good plan for those utilizing self-custody. First, your bitcoin units don't need to sit on an exchange for any amount of time whatsoever. Use the exchange only to make purchases. After your bitcoin purchase is completed, move the private keys to self-custody.

Since cold storage is the preferred method for long-term HODLers, and the safest storage method for large amounts of money, you should aim to put most of your bitcoin wealth into cold storage. I only advise putting a very small portion of your overall bitcoin funds in a hot wallet, simply for the sake of liquidity.

At the end of the day, you want peace of mind, to build and grow your bitcoin investment over many years without worrying about it. Self-custody, cold storage, and physical protection for your cold storage and seed phrase gives you beautiful peace of mind.

* * *

It's imperative to take the time to research your options and devise a bitcoin storage strategy that best suits you. While it's solely your responsibility, it's also a very fun part of Bitcoin. Don't be intimidated. You'll learn more along the way. Don't be paralyzed by information and just get started. I assure you that it'll all become second nature to you soon enough.

With that said, let's move on to the final chapter in *The Bitcoin Shortcut*. It covers some very important things you need to think about and decisions you should make before investing your money in Bitcoin.

CHAPTER 17

# BITCOIN INVESTMENT STRATEGY

This chapter contains my opinions about investing in Bitcoin, specifically how I think you should go about it, things you should consider, and things you should avoid. I want to make two things clear before we get started.

The first is that your financial situation, investments, and the risks you take with them are your responsibilities—not mine, not the government's, not your neighbor's, not anyone else's. What you do or don't do with your money is on your shoulders.

I refer you to the copyright page of this book for relevant legal disclaimers. Basically: if *you* do something stupid with *your* money, it's not *my* fault. Perform your own due diligence, research, and make your own informed decisions. My opinions in this chapter are meant to guide you and get you thinking, they are not edicts.

Second, I'm not some crypto fanboy who just pulled some investment advice out of hat. Most of my life has been involved in finance and investments. To give you a summary, it started when I earned my university degree in finance. I then spent over a decade in wealth management for ultra-high-net-worth individuals and institutions.

In that capacity, I mainly performed investment analysis, investment selection, and portfolio management. On a daily basis I dealt with currencies, stocks, bonds, mutual funds, ETFs, options, commodities, futures, hedge funds, private equity, real estate, and just about everything else you can invest in.

During that time, I also became a Chartered Financial Analyst charterholder. The CFA designation is a postgraduate professional certification covering a variety of investment topics. It was a much more valuable and challenging experience than my undergraduate degree.

This chapter (and Chapter 14) are well informed. I certainly wouldn't try to give advice on something I don't know about.

So, with that in mind let's get started. You already know that Bitcoin is a revolutionary technology. As an investment, it deserves to be treated entirely differently than any other investment available to you.

Since the nature of Bitcoin requires more responsibility for your actions, I think it is fitting that you find your own answers to the important Bitcoin investment questions and implement your own investment plan. Have no fear—investing in Bitcoin is not scary or complicated. You can do it and have a lot of fun at the same time.

## NOTES ON FINANCIAL ADVISORS

Financial advisors live in a world of well diversified, multi-asset portfolios that are analyzed and built using things like beta, correlation, Sharpe ratio, standard deviation, asset class weightings, asset location, and so on. I have nothing against the smart, competent ones that *actually* care about their clients (they are very rare). The bad ones I despise.

Good advisors have an important role for clients regarding traditional investments and traditional portfolios, no doubt about that. They really can help you with a lot related to your finances.

However, you don't need a financial advisor for *anything* related to your Bitcoin investment.

I can hear the financial advisors of the world screaming at me right now, belching out a million reasons why I'm wrong. That's OK. Wall Street will never agree with me on this. Financial institutions will try very hard to make money from your Bitcoin investment by claiming you need their "comprehensive portfolio advice" or "proprietary Bitcoin products." That's drivel at best.

Stay on guard. See through their constant attempts to convince you that they're needed for Bitcoin. They aren't.

Why don't you need a financial advisor for your Bitcoin investment?

It's quite simple really. Bitcoin is a revolutionary long-term store of value. There is enormous amount of potential financial return to be had over the coming years and decades as a tsunami of money flows into this asset. The world needs Bitcoin, the world wants Bitcoin, and the world is rapidly investing in Bitcoin.

Make no mistake, *everything* to do with money will become decentralized this century. Banking, payments, investments, insurance, you name it. It's *all* going to shift to a decentralized peer-to-peer model. And through this entire financial evolution Bitcoin's has a foundational role to play. It will be the world's standard, base-layer store of value that the entire global decentralized monetary system uses as a final settlement mechanism. (Pay attention to what smart money and smart companies are already doing with Bitcoin. This endpoint is obvious.)

With all of this in mind, you just need to answer three questions:

1. Do you *want* to invest in a revolutionary long-term store of value?
2. If yes, *how much* money do you want to invest?
3. *How* should you go about investing that amount?

You're an adult. You can answer these questions yourself. You don't need anyone else to tell you the answers or try to influence your answers. Thinking for yourself is a beautiful thing.

Regarding the first question. I sure hope this book has convinced you what a revolutionary technology Bitcoin is and how it's a once in a lifetime opportunity. I hope for the sake of your financial future that you said yes. But you know yourself better than anyone else. Only *you* can answer that question.

The rest of this chapter covers questions two and three. When you answer these three questions you won't need a financial advisor for your Bitcoin investment.

**There is never a reason to pay any fees to a financial advisor for your Bitcoin investment.**

Financial advisors are not bad people (well, some are vile commission-thirsty product pushers who deserve one-way tickets into orbit, but that's another book). The good ones are just normal people trying to earn a living just like everyone else. They usually make their money either by flat fees for advice or by annual percentage fees of your investment assets.

Some even also use performance-based fees, taking a percentage of your profits when you do good but, curiously, not paying you any money when investments go down in value. This risk asymmetry is a lack of skin in the game that always enraged me and is one of the reasons I left that industry for software. (They've come up with ways to trick you into thinking this asymmetry doesn't exist. One such idea is called a high-water mark. This doesn't address the asymmetry at all.)

Anyways, let's look at each common fee type and why you shouldn't pay it for your Bitcoin investment.

*Don't pay flat fees for advice.* You've made the decision to invest in Bitcoin. You'll learn how to determine how much money to invest and how to invest that money in this chapter. Your three questions will be answered. You don't need any advice.

*Don't pay an annual management fee on your Bitcoin balance.* Bitcoin is a store of value that just sits there. No major decisions need to be made about it. You're going to accumulate Bitcoin and HODL for years. So why would you pay an annual management fee on this balance? Makes zero sense.

There's nothing to advise on—recall it's a non-overlapping asset that you've mentally and physically separated from your traditional investments. If financial advisors attempt to use all their bunk Modern Portfolio Theory tools on a blended portfolio of traditional investments and Bitcoin, both your traditional investments and your Bitcoin balance will suffer. You'll lose out twice. Leave Bitcoin and your other investments separate.

This is an enormous, jagged, irregularly shaped, hard pill for financial advisors to swallow since it goes against everything they know and have been trained to do. There's no way an honest financial advisor can justify charging you a fee on *your* Bitcoin keys stored in *your* cold wallet that's hidden somewhere in *your* house, especially if that advisor truly understands how Bitcoin works.

*Don't pay commissions on purchases.* Never pay a commission to an advisor for a Bitcoin investment. You'll be doing the actual trades yourself so why pay a commission to anyone else?

Also, it's not a good idea *at all* to give your financial advisor access to your exchange account or wallet—don't be foolish. Since they won't have access, they won't be "doing the trade for you." And don't let them sucker you into the idea of them buying the bitcoin first and then sending it to your wallet. You can make the purchases yourself directly. The concept of commissions is a poisonous barnacle on the entirety of finance. Commission needs to go. Never, ever pay a commission. If an advisor is not 100% transparent about commissions (and their entire fee structure), they are not a nice human being. Period.

*Don't pay performance-based fees.* Are you going let Advisor Sally take a percentage of your profits on Bitcoin just for telling you to invest in Bitcoin? Yet, if you lose money, she won't pay you for going under water. It's silly. Run away if an advisor suggests this fee structure to you. There is no situation ever where a financial professional should earn a percentage of your Bitcoin profits. Why let them piggyback with no consequences?

Be vigilant. Advisors will try to sell you on the idea that they can help you determine how much money to invest in Bitcoin and that they can help "periodically rebalance your portfolio" to manage risk. Believe me they'll try hard.

As mentioned, you'll be able to determine how much money to invest yourself. *Is it really that difficult to do?* As for the concept of rebalancing, that makes a lot of sense with traditional investments, and even alternative ones, but it's **meaningless** with Bitcoin. The goal is to acquire as much Bitcoin as possible and HODL for a decade or more. This implies that there will be no selling and no rebalancing (another hard pill for financial advisors to swallow). You can worry about selling many years down the road.

**Here's the cold hard truth**: if a financial advisor didn't read this book, then you know Bitcoin better than he or she does. Keep that in mind.

Certainly, allow your financial advisor to advise on, adjust, rebalance, and optimize your traditional investment portfolio. Honest, competent, smart advisors do have value to add there and can probably justify their fees. But you can handle your Bitcoin investment all by yourself—it isn't rocket science. Remember, Bitcoin is a non-overlapping asset. You have your traditional investments in one bucket, and you have your Bitcoin, proudly standing alone, separate from the system.

Regarding the three questions mentioned early in this section. Keep your answers in mind and at the ready and you'll do just fine.

## INVESTING CONSIDERATIONS

By now you're probably wondering "how much should I invest in Bitcoin?" The answer to this question is something that you must think through carefully. Luckily there is a useful framework that can help. It's a seven-letter initialism: *RRLLTTU.*

RRLLTTU is used often by financial professionals to help clients construct a traditional investment plan and a well-diversified portfolio of assets to help them reach their goals. Great.

Since I view Bitcoin as a non-overlapping asset, I obviously *don't* think it should be viewed in the typical "well-diversified portfolio of assets" way. Bitcoin should be bucketed alone, separated—mentally, physically, emotionally, and financially. This means that you should utilize RRLLTTU for Bitcoin as *a standalone investment.*

Don't think about any of your other investments here. Focus on Bitcoin and Bitcoin only. It's your choice to view Bitcoin as a non-overlapping asset or not, but I'll continue the rest of this chapter assuming that you do.

RRLLTTU stands for the seven factors you should consider when coming up with a comfortable amount to invest in Bitcoin. They are: risk, return, liquidity, legal, taxes, time horizon, and unique circumstances.

Would you put 99% of your net worth in Bitcoin? Some people have. Would you only want to put less than 1% of your net worth in Bitcoin? Some people do just that. Is $1,000 week a good amount to start investing, or is $20 a week the better option for you?

RRLLTU can help you decide these things. And decide you must. **No one else can tell you how much to invest in Bitcoin.** It's your choice. I'll lay out some broad strokes below, but you need to spend time thinking about what's best for you. Let's discuss each factor in more detail.

**Risk** - How much risk can you handle? How much risk are you willing to take? Your risk tolerance is broken up into two components: *ability to take risk* and *willingness to take risk.* Your ability to take risk refers to your actual money situation. Do you have excess income? Is your cash flow good? Are your liabilities under control? Do you have enough cash savings for emergencies? Your willingness to take risk refers to your desire—do you *want* to take risk with your cash? Are you interested in putting some safe cash at risk in a volatile investment?

If you have a high ability and a high willingness to take risk, you can be more comfortable investing larger amounts in Bitcoin. If you have low ability to take risk coupled with a low willingness, you need to start very small.

There's a broad range of possibilities between these two components of risk tolerance. Identify where you stand. You shouldn't invest $500 into Bitcoin if you know you need that $500 in a week. During the week, your investment may decrease to $250 and that wouldn't be good for you. But hey, it might be worth $1,000.

**Know your risk tolerance and only invest money that you can afford to live without. In 10 years, it may be worth much more or it could be worth very little, maybe even zero.**

Note that Bitcoin is a very risky asset for the time being. It's very volatile with big price swings, but if you're a HODLer, that won't matter. Don't think of Bitcoin as something you trade in and out of regularly; you'll lose your shirt trying to time the market.

**Return** - It the traditional context with a multi-asset class diversified portfolio, calculating your "required return" or "desired return" might be a useful exercise. With Bitcoin, it's not. Your goal should be to HODL as long as possible. You won't be sucking on the teat of your Bitcoin portfolio for sustenance over the next decade—that's what your fiat income is for. So don't worry about required or desired return; it's a pointless exercise with Bitcoin.

**Liquidity** - With traditional investments, taking liquidity into consideration is very important. Do you need large chunks of money at a certain point in time? Should you be constrained from investing in assets that have little or no liquidity in the marketplace? These are the types of questions that help determine your plan.

Bitcoin is very liquid so *if* there comes a time soon that you absolutely need those funds, you don't have to worry about it being locked up or hard to get. You can trade out of Bitcoin on a moment's notice.

At the same time, if you know that you have a big cash outflow coming in the near future, such as a down payment on a house, you probably don't want to keep those funds in Bitcoin. There's too much short-term risk and you may not be able to meet your obligation.

Handle liquidity issue and concerns with your traditional portfolio and your income. Only after your liquidity issues are solved, met, and planned for should you invest money in Bitcoin. At the same time once you are invested in Bitcoin, try not to rely on those funds for liquidity. HODL.

**Legal** - Are there any legal constraints or considerations that you need to take in account? Different countries deal with Bitcoin in different ways. Even jurisdictions within a particular country may handle Bitcoin differently. You must do your homework to determine if there's anything special regarding your personal situation that you need to consider before investing.

On the very far end of the spectrum, ask yourself: Would I move to a more Bitcoin-friendly country if needed to protect my investment and avoid legal trouble? Start planning those contingencies. It's not too far out there; the world is changing fast. Be prepared.

**Taxes** - Every jurisdiction in the world has its own tax treatment of Bitcoin. You must do your homework here, too. What are the general tax considerations you must think about? What are your specific tax considerations? Is there anything special about your tax situation that could or should change your Bitcoin investment plan?

In the U.S. since 2014 the law has been clear: Bitcoin is an asset, not a currency. Any net gain from selling or transacting in Bitcoin is subject to capital gains taxes. So, if you're in the U.S. it'd be helpful to learn what short-term capital gains, long-term capital gains, and marginal income tax rates are. You just need to understand the broad strokes of these ideas. You don't need to become a tax expert.

**Time Horizon** - You should view Bitcoin as a long-term investment, looking out 10 years or more. However, there may be special factors that might affect your specific Bitcoin investment plan's timeframe. Should you view your time horizon differently than the typical long-term Bitcoin investor? Perhaps, but only you can decide that. Are you in this for the long haul, or just a few months? One to five years? Or a few decades?

**Unique Circumstances** - Everybody has some sort of unique circumstances that will affect their investment plan. Think through them and act accordingly. It's most helpful to view these unique items in the context that Bitcoin is a long-term investment that you'll HODL. It's also useful to remember that Bitcoin is an asset that's highly volatile for the time being. If you know you need to pay for grandma's hip replacement surgery in 7 months, don't put those funds in Bitcoin.

There's no magical formula for determining how much money to invest in Bitcoin or how often. The only useful way to arrive at these numbers is to think through as many factors as possible. Don't take on more risk than you are comfortable with and keep educating yourself on Bitcoin. A fool and his money are soon parted. Don't be a fool.

## DOLLAR-COST AVERAGING

After you've taken into consideration RRLLTTU there's only one actual investment implementation strategy that you need. It's called *Dollar-Cost Averaging (DCA).* In simple terms it means that you decide how much money you want to invest and how often. Then you stick with that same dollar amount and that frequency, investing like clockwork *regardless of the price of Bitcoin.*

With DCA you don't try to time the market, ever. You simply pull the trigger and make the purchase at the predetermined time with the predetermined amount of money. And you never miss an investment day.

For example, say you get paid on the second and fourth Friday of each month. You've decided that you're going to put $100 of your net paycheck into Bitcoin each paycheck. So come hell or high water, when that paycheck hits your bank account on the second Friday of the month, you buy $100 of Bitcoin. When the next paycheck hits on the fourth Friday of the month, you buy $100 of Bitcoin.

It doesn't matter if the price is up or down. It doesn't matter at all what the price of Bitcoin even is that day. You don't need to worry about the price at all. You just make the purchase. $100 every two weeks. Don't stress over the exact timing of the purchase—just make the purchase ASAP that day.

And you do the same thing the next month, and the next, and the next. You keep following this DCA schedule until you make a conscious decision to adjust your Bitcoin investment plan.

Usually, people who use the DCA strategy make one or two purchases per month. But if you want to DCA every single week, or once every three months, it's all up to you. Just choose a *frequency of purchases* and the *amount of money* you want to invest *each time period.* Then just stick with the plan. If you miss an investment day, it's not the end of the world, just get back on track as soon as you can. If you forget, just invest the next day. If you don't have the money, don't stress out. Just invest some cash the next time you get some.

DCA is a very plain-vanilla strategy but it's the perfect strategy for Bitcoin. You don't need any fancy tools, analytics, information, mentors, or advisors—you just need discipline. You'll wind up buying more bitcoin quantity when the price is low, and less when the price is high. But steadily you will build your quantity of satoshis (and eventually bitcoin). Getting rich through Bitcoin is an ultramarathon. Stay in the race—it's many years long.

DCA helps lessen the impact of volatility, reduces stress, and gives you peace of mind. Don't chase short-term returns on Bitcoin—focus on building your quantity of bitcoin over the long-term. The financial rewards will come in a decade. Be patient and you'll be thankful.

## DAY TRADING BITCOIN

Don't day trade Bitcoin. Instead, Dollar-Cost Average and HODL. Day trading = market timing = losses. You might have some good days, but I'd bet my bitcoin that over the long run you'll come out with less money than you started with.

How do you make a small fortune in Bitcoin? Start with a large one and day trade it.

## IGNORE VOLATILITY

Anyone investing in Bitcoin needs to realize that it's a highly volatile asset for the time being. This means that the value of it swings wildly. It's quite common to see double-digit returns in a 24-hour period, both on the upside and the downside.

You may invest now and a few months later see that your money has doubled. But it's also very possible that you'll see it be cut in half.

The volatility of Bitcoin will stabilize over the coming years, but for now its value changes a lot week by week and day by day. You'll sometimes even see large price swings in the period of a single hour.

**You must always keep in mind that Bitcoin is a long-term investment. You're in it for the long haul: 10, 15, 25 years or more. Don't pay attention to the short-term volatility. Just ride the wave. The daily price noise means zero to long-term investors.**

If you're the type of person to put $100 into an investment and watch it like a hawk every few minutes, stressing out if it goes down to $85, then up to $110, and back down to $97, then Bitcoin may not be for you at this time. You probably need a vacation.

## LEGACY PLANNING

This part of a person's Bitcoin investment plan is often overlooked. Legacy planning is the process of preparing how to bequeath your assets to loved ones after you die.

If you don't have a plan in place your bitcoin will likely go into limbo forever. Wouldn't you rather have your spouse, children, or even a friend get that money after you're gone?

I'll give you three tips on bitcoin legacy planning.

The first tip is to avoid the trap of thinking bitcoin legacy planning isn't for you. It is. Your loved ones or next of kin need to know how to access your bitcoin *when the time is right* so they can have it. There's no Bitcoin Fairy that will magically appear and put your bitcoin keys under their pillows when you die.

You don't have to give them access to your private keys right now, but you should have a plan in place, so they'll know what to do when you die. This includes if you die suddenly and unexpectedly. Take this seriously and plan so your bitcoin doesn't die along with you.

The second tip is to **never** put private keys in a last will and testament. That leaves your bitcoin wide open for theft because wills become public record. Here's a quote from the New York State Court's website:

> A Will is a confidential document until the person dies. Once the Will is admitted to probate or a small estate, it becomes a public record document that anyone can see and read.

This is bad for your bitcoin for obvious reasons. Don't put your keys in a will otherwise someone may beat your heirs to that money.

The third tip will help you figure all of this out. There is a great little book titled *Cryptoasset Inheritance Planning* by Pamela Morgan, Esq. that you should buy, read, and work through. There isn't much else to say about that. You just need to do it.

* * *

There you have it, a simple and direct plan for investing in Bitcoin. Though certain parts take some thought, it's all very straightforward. **The key is having discipline to do it.**

You don't want to look back in 10 years when Bitcoin is likely valued at 7-figures a coin and think, "damn it, why didn't I *do* this?" Ten years from now will surely arrive. Will you be wealthy thanks to Bitcoin? I sure hope so.

Get started today. Get started now. You know exactly what to do.

# CLOSING THOUGHTS

You've come a long way. At first you may have been skeptical about Bitcoin but now I hope that you've seen the light.

Bitcoin is changing the world and there's no stopping it. Now you can safely say that you *really* understand what Bitcoin is and exactly how it works. You can describe its potential and you're aware of the massive financial opportunity that lies ahead.

Hopefully you start your Bitcoin investment journey today. Build your nest-egg of bitcoin over the coming years and don't regret missing the boat. Bitcoin can make all the difference for you.

Keep learning more about Bitcoin and the cryptocurrency ecosystem. Help others along the way. The more people that get involved in Bitcoin the better the world will be.

I sincerely hope that Bitcoin changes your life as much as it has changed mine.

Craig Anthony, CFA

www.craiganthony.io

@craiganthony_io (Twitter/X)

## APPENDIX 1

# SHA256

Since Bitcoin relies on the SHA256 cryptographic hash algorithm, I thought it would be helpful to explain it a bit further. Though I won't provide a full example since that would require lots of computer code, I can walk you through the basic steps and major ideas of the algorithm.

Though you already know this by now, I just wanted to remind you that this is a one-way function. There are no known ways to take an output hash, reverse these steps, and arrive at the input. For a refresher on what cryptographic hash functions are please revisit Chapter 6.

SHA256 does the following:

- Takes an input of any size
- Scrambles the data deterministically and irreversibly
- Outputs a 256-bit hash

The middle bullet point is what we'll focus on. It consists of eight steps:

1. Pre-process the data
2. Initialize hash values
3. Initialize round constants
4. Perform loop on "chunks" of data

5. Create message schedule
6. Compress
7. Modify values
8. Concatenate the final output hash

Let's briefly look at each of these eight steps.

**Step 1** - This is the pre-processing phase. First take the input and convert it to binary (0's and 1's). Then append a single 1 to the end of this binary data. Next, pad the data with 0's until the data is a multiple of 512, less 64 bits. Lastly, append a special 64 bits to the end. These 64 bits are a big-endian integer representing the length of the original input in binary. *The result of this pre-processing will always be evenly divisible by 512.*

**Step 2** - Here, eight hard-coded hash values are created. They are used every single time something is hashed with SHA256. These eight hashes represent the first 32 bits of the fractional parts of the square roots of the first eight prime numbers. These eight variables, `h0` to `h7`, are initially set to the following values:

```
h0 = 0x6a09e667
h1 = 0xbb67ae85
h2 = 0x3c6ef372
h3 = 0xa54ff53a
h4 = 0x510e527f
h5 = 0x9b05688c
h6 = 0x1f83d9ab
h7 = 0x5be0cd19
```

**Step 3** - Here 64 more constants, represent by `k0` through `k63`, are created. These are hard coded, just like in

Step 2. These 64 constants are the first 32 bits of the fractional parts of the cube roots of the first 64 primes. For simplicity, I will not list `k0` through `k63`, but they look similar in style to `h0` through `h7` listed in Step 2.

**Step 4** - In this step, for those that know programming, a loop is performed. The loop performs the remaining steps on each 512-bit chunk of data from Step 1. Each iteration of the loop mutates the values of `h0` through `h7`.

*For steps 5, 6, and 7, all addition is calculated modulo $2^{32}$*

**Step 5** - Take the result of Step 1 and break in into 32-bit pieces. Each of these pieces is put into an array. A 512-bit chunk has 16 pieces that are 32 bits each. At the end of this array, add 48 more 32-bit pieces, each initialized to zero. We now have an array, *w*, of 64 pieces: *w = [0…63]*.

Pieces of this array are then manipulated in an algorithm that looks like this:

```
for i in w[16...63] {

  s0 = (w[i-15] rightrotate 7) xor (w[i-15]
  rightrotate 18) xor (w[i-15] rightshift 3)

  s1 = (w[i-2] rightrotate 17) xor (w[i-2]
  rightrotate 19) xor (w[i-2] rightshift 10)

  w[i] = w[i-16] + s0 + w[i-7] + s1

}
```

**Step 6** - Next a compression loop is used. Eight variables are created: `a`, `b`, `c`, `d`, `e`, `f`, `g`, `h`. They are initially set to:

```
a = h0
b = h1
```

```
c = h2
d = h3
e = h4
f = h5
g = h6
h = h7
```

These new variables will be mutated during each pass of the loop.

```
for i in [0…63] {

  s1 = (e rightrotate 6) xor (e rightrotate
  11) xor (e rightrotate 25)

  ch = (e and f) xor ((not e) and g)

  temp1 = h + s1 + ch + k[i] + w[i]

  s0 = (a rightrotate 2) xor (a rightrotate
  13) xor (a rightrotate 22)

  maj = (a and b) xor (a and c) xor (b and
  c)

  temp2 := s0 + maj
  h = g
  g = f
  f = e
  e = d + temp1
  d = c
  c = b
  b = a
  a = temp1 + temp2

}
```

**Step 7** - The last step of the loop on the "chunks" of data is as follows:

```
h0 = h0 + a
h1 = h1 + b
h2 = h2 + c
h3 = h3 + d
h4 = h4 + e
h5 = h5 + f
h6 = h6 + g
h7 = h7 + h
```

**Step 8** - The very last step is to concatenate `h0` through `h7` together. In this step, the plus sign means "append the value to the end of the previous value".

Hash = `h0` + `h1` + `h2` + `h3` + `h4` + `h5` + `h6` + `h7`

* * *

And there you have it. A brief description of how the SHA256 cryptographic hash algorithm works. Full code examples can be found online.

# APPENDIX 2

# ENTROPY

The word entropy comes from thermodynamics and is a measure of the amount of a system's thermal energy that is unavailable to be converted into mechanical work. It's interpreted as the degree of disorder or randomness in the system.

In the information security sense *password entropy* is the measure of how unpredictable a password is. The higher the entropy the better for security and the more difficult is it to correctly guess the password.

Password entropy has a simple formula:

$$E = \log_2(R^L)$$

where,

E = the number of bits of entropy
$R$ = number of unique characters available
$L$ = number of characters used in the password

$R^L$ is equal to the number of possible passwords given $R$ and $L$. The password entropy grows with increases in $R$ and $L$.

Computers speak in the language of 0's and 1's. When there is a choice between a single 0 or 1, this is the same as a coin flip. With a single coin flip, $R$ is equal to 2 since the possibility of heads or tails exists. And $L$ is equal to 1 since a toss is a single toss with only one outcome. Assuming a totally fair coin, a single coin toss represents 1 bit of entropy:

$$E = \log_2(2^1) = 1$$

Entropy is a measure of uncertainty. If you lower entropy, you lower uncertainty and gain information. If you increase entropy, you lose information, and your uncertainty increases. Take a fair and normal 6-sided die. The values of its sides are {1,2,3,4,5,6}. Each of these values is equally likely to come up on a roll with a $1/6^{th}$ chance. The entropy of a 6-sided die is 2.585:

$$E = \log_2(6^1) = 2.585$$

What happens if we double the number of sides? Well, if we have a 12-sided die we double the number of possible results. The values now are {1,2,3,4,5,6,7,8,9,10,11,12}. The entropy of this die is 3.585:

$$E = \log_2(12^1) = 3.585$$

For every doubling of the number of sides the entropy increases by exactly 1 bit. If we go from a 12-sided die to a 24-sided die, the entropy increases to 4.585. A 48-sided die has 5.585 bits of entropy, and so on.

Let's go back to the 6-sided die. Imagine that we double the *values* on each side so our value set is now {2,4,6,8,10,12}. Does the entropy increase? No. The entropy of this die would also be 2.585. This is because each of these values *still has a $1/6^{th}$ probability of occurring*. For a die, the number of sides affects its entropy, not the values of the sides.

More generally, the entropy of a discrete variable depends on the probability distribution of its outcomes, not the values of those outcomes.

Now that you have a little intuition about what a single bit of entropy means, let's take the lowercase English alphabet:

```
abcdefghijklmnopqrstuvwxyz
```

This has 26 unique characters available. Let's look at the amount of entropy for this character set when choosing a 5- and 10-character password:

$$E = \log_2(26^5) = 23.5$$

$$E = \log_2(26^{10}) = 47.0$$

Let's see what happens to 5- and 10-character password entropy when *R* is upgraded to be all lowercase and uppercase English letters, plus the ten numeric characters:

```
abcdefghijklmnopqrstuvwxyzABCDE
FGHIJKLMNOPQRSTUVWXYZ0123456789
```

$$E = \log_2(62^5) = 29.7$$

$$E = \log_2(62^{10}) = 59.5$$

If we then add ten special characters, such as `/*~+!@#$%&`, our ten character password's entropy is almost 62. Entropy is the exact reason that many online accounts require that passwords be a certain length and contain a variety of characters. To achieve a password entropy of 62 with only lowercase English letters would require that the password be 14 characters long (and it would be 65.8 bits of entropy).

Any entropy less than 36 is considered weak. Up to 60 is reasonable, and secure enough for simple accounts. From 61 to 127 bits of entropy, the strength is good enough to protect more sensitive information, even financial information. Entropy that is 128 bits or higher is considered very strong and even more than needed, perhaps overkill.

Recall that Bitcoin Hierarchical Deterministic wallets utilize a 12- or 24-word seed phrase to generate your private key tree structure. A 12-word seed phrase represents 132 bits of entropy, while a 24-word seed phrase represents 264 bits of entropy.

For a 12-word phrase, a random sequence of 128 binary digits is generated, then the 4-bit checksum is added, resulting a random sequence of 132 binary digits (0's and 1's).

$$E = \log_2(2^{132}) = 132$$

A 24-word seed phrase has an 8-bit checksum added, resulting in a random sequence of 264 binary digits.

$$E = \log_2(2^{264}) = 264$$

A 12-word seed phrase protecting your private keys is extremely strong and a 24-word seed phrase guarding your private keys is *ridiculously* strong. Some people think that backing up a 24-word seed phrase is a hassle compared to a 12-word seed phrase. Now that you know more about entropy, I'll let you be the judge.

Refer to Chapter 10 for a refresher on wallets and seed phrases.

# SELECT REFERENCES

During the process of researching and writing *The Bitcoin Shortcut* I read and referenced more physical books than I can count. The following list makes the cut as recommended reading.

**General, intuition building books about Bitcoin and related topics**

Timi Ajiboye, et al., *The Little Bitcoin Book* (Whispering Candle, 2019) ISBN 978-1-64199-050-9

Andreas M. Antonopoulos, *The Internet of Money Volume One* (Merkle Bloom LLC, 2016) ISBN 978-1537000459

Andreas M. Antonopoulos, *The Internet of Money Volume Two* (Merkle Bloom LLC, 2017) ISBN 978-1947910065

Andreas M. Antonopoulos, *The Internet of Money Volume Three* (Merkle Bloom LLC, 2019) ISBN 978-1947910171

Saifedean Ammous, *The Bitcoin Standard* (John Wiley & Sons, 2018) ISBN 978-1119473862

Nik Bhatia, *Layered Money* (Independently published, 2021) ISBN 978-1-7361105-2-2

Kiara Bickers, *Bitcoin Clarity* (Bickers & Son, 2020) ISBN 978-1-7338712-0-4

Chris Burniske and Jack Tatar, *Cryptoassets* (McGraw-Hill Education, 2018) ISBN 978-1-260-02667-2

Phil Champagne, *The Book of Satoshi* (E53 Publishing, 2014) ISBN 978-0-9960613-1-5

Pamela Morgan, *Cryptoasset Inheritance Planning* (Merkle Bloom LLC, 2018) ISBN 978-1-947910-11-9

Yan Pritzker, *Inventing Bitcoin* (Independently published, 2019) ISBN 978-1794326316

## Technical books about Bitcoin and related topics

Andreas M. Antonopoulos, *Mastering Bitcoin* (O'Reilly Media, 2017) ISBN 978-1-491-95438-6

Imran Bashir, *Mastering Blockchain* (Packt Publishing, 2020) ISBN 978-1-83921-319-9

Pavel Kravchenko, et al., *Blockchain and Decentralized Systems Volume 1* (Distributed Lab, 2018) ISBN 978-6-1776-3428-6

Pavel Kravchenko, et al., *Blockchain and Decentralized Systems Volume 2* (Distributed Lab, 2020) ISBN 979-863-7553-55-6

Pavel Kravchenko, et al., *Blockchain and Decentralized Systems Volume 3* (Distributed Lab, 2020) ISBN 978-617-7634-79-8

Lorne Lantz and Daniel Cawrey, *Mastering Blockchain* (O'Reilly Media, 2021) ISBN 978-1-492-05470-2

Arvind Narayanan, et al., *Bitcoin and Cryptocurrency Technologies* (Princeton University Press, 2016) ISBN 978-0-691-17169-2

Kalle Rosenbaum, *Grokking Bitcoin* (Manning Publications, 2019) ISBN 978-1617294648

Jimmy Song, *Programming Bitcoin* (O'Reilly Media, 2019) ISBN 978-1-492-03149-9

## Other books you should read

James Dale Davidson and Lord William Rees-Mogg, *The Sovereign Individual* (Touchstone, 2020) ISBN 978-0-684-83272-2

James V. Stone, *Information Theory* (Sebtel Press, 2015) ISBN 978-0-9563728-5-7

Nassim Nicholas Taleb, *Antifragile* (Random House Trade Paperback, 2014) ISBN 978-0-8129-7968-8

## Extra goodie

I included this book for anyone interested in the programming language that Satoshi used to code Bitcoin, C++. It's a great reference.

Josh Lospinoso, *C++ Crash Course* (No Starch Press, 2019) ISBN 978-1593278885

If you're even the slightest bit interested in programming, give it a shot, and start today—just pick a language and go. The skillset makes you unstoppable.

# THANK YOU

Thank you for reading *The Bitcoin Shortcut*. I hope you learned a lot and agree that Bitcoin is incredible.

If you liked the book, please help by giving it a 5-star rating on Amazon. Also please consider writing a short positive review on Amazon—they mean so much to me.

Please also tell your friends about the book, especially those who are excited about Bitcoin. And don't forget that *The Bitcoin Shortcut* makes a great gift for people you care about.

I really appreciate the ratings, reviews, and shares. Thank you!

**www.TheBitcoinShortcut.com**

# ERRATA

No matter how many times the draft of a manuscript is read and edited, there are always grammatical errors and typos lurking in the pages.

If you find an error, please reach out to me on Twitter and I'll correct the mistake in the next update of the book.

Thank you for your help.

Significant price discounts are available for corporate or personal orders of 100 books or more. Please go to www.craiganthony.io and contact Craig directly.

**www.craiganthony.io**

Go to www.craiganthony.io to learn more about Bitcoin from Craig Anthony, CFA.

# HODL

Made in United States
North Haven, CT
13 January 2025

64377464R00157